Urban Climate Change Adaptation in Developing Countries

Mohsen M. Aboulnaga • Amr F. Elwan
Mohamed R. Elsharouny

Urban Climate Change Adaptation in Developing Countries

Policies, Projects, and Scenarios

Springer

Mohsen M. Aboulnaga
Department of Architecture, Faculty of Engineering
Cairo University
Giza, Egypt

Amr F. Elwan
Faculty of Engineering
Military Technical College
Cairo, Egypt

Mohamed R. Elsharouny
Architect & Out-sourced Project Coordinator
African Export Import Bank (Afreximbank)
Cairo, Egypt

ISBN 978-3-030-05404-5 ISBN 978-3-030-05405-2 (eBook)
https://doi.org/10.1007/978-3-030-05405-2

Library of Congress Control Number: 2018968019

This Springer imprint is published by the registered company Springer Nature Switzerland AG
The registered company address is: Gewerbestrasse 11, 6330 Cham, Switzerland

Preface

Nothing is permanent! We all know that climate change is inevitable; nevertheless, we are alarmed by the changes occurring in the global climate. Currently, the prime concern that future generations face is the occurrence of undesired climate change risks, which have been manifested worldwide. Therefore, the authors have written this book to focus on understanding climate change impacts on the built environment and related sectors, and to explore solutions for these problems through the implementation of climate change adaptation policies, measures and actions as well as programmes and projects. One of the challenges highlighted by the literature in this book, is that adopting policies is not enough to counterbalance climate change, nevertheless, implementing actions and tools such as adaptation decision frameworks are imperative to move adaptation from theory to practice and to overcome obstacles and guarantee successful adaptation to climate change.

The book's contents are structured into two main parts and are presented using a farmed approach to assist the reader in facilitating and comprehending the contents. The first part of the book presents the global risks of climate change and investigates the consequences on the built environment and its impact on different sectors such as agricultural and food security, ecosystems, forests, water, health, coastal and flood prone zones and tourism and energy. While the second part highlights the climate change measures, policies and actions at the global scale, it focuses specifically on climate change adaptation in developing countries. In addition, the lead by example approach is addressed in Part II by presenting nine climate change adaptation projects in one of the developing countries – Egypt, and comparing them to global projects in other developing countries, with the purpose in mind to identify the gaps and track success stories, and failures in the climate change adaptation process.

With contributions from the lead author, an international and national expert who has more than 30 years of experience in higher education, government, senior management and consultancy in sustainable energy policies and climate change adaptation, this book will be a great resource on the topic of 'climate change'. We strongly believe that the book will interest governments, professionals, research centres, libraries, researchers and students who are interested in this field, as well as policy-makers and industry stakeholders involved in climate change adaptation and mitigation.

You are most welcome to contact the lead author at, mohsen_aboulnaga@yahoo.com, to share your thoughts and comments, while you are reading this book or afterwards. I promise to get back to you in a timely manner.

Cairo, Egypt Mohsen M. Aboulnaga

Acknowledgements

I would like to express my sincere thanks to the co-authors who have contributed tirelessly and significantly to make this book a reality, and for providing their thoughts, research work and wealth of materials that have contributed to this volume.

We would like to thank the Springer editorial team and production team for their valuable advice, support and patience during the development of this book. I would like to thank Amanda Quinn, Faith Pilacik, Brian Halm and Abhishek Ravishankar for their kind efforts and endless assistance during the process.

Special thanks go to Mona Mostafa for her amazing efforts and continuous assistance during the writing and development of this book. We also thank Perihane Wael for her support and time taken to proof the book chapters.

Finally, our thanks are extended to everyone who has helped bring this book to completion.

Mohsen M. Aboulnaga
Amr F. Elwan
Mohamed R. Elsharouny

Contents

About the Authors

Mohsen M. Aboulnaga has more than 30 years of experience in higher education, government, senior management and consultancy in sustainable energy, including energy efficiency and renewable energy, energy policy development and green cities as well as climate change adaptation and academic positions. Dr. Aboulnaga is a Professor of Sustainable Built Environment at the Faculty of Engineering, Cairo University, and Government Strategy and Policy Advisor. He is also an Associate and Senior Consultant – Strategy and Sustainable Development at Oil and Gas Skills (OGS) – at the Human Development Arm of the Ministry of Petroleum and an Associate Expert at the Regional Centre for Renewable Energy and Energy Efficiency (RCREEE) where he represented it at the high-level meeting at the European Commission Joint Research Centre (EC-JRC) in Ispra, Italy, to review the Guidebook for ENPI South countries. He is an Expert on sustainability, sustainable development and sustainable urban development, including energy efficiency, green building and green and smart cities in developing countries, Gulf region and worldwide. His areas of competency include strategy planning, policy development, low-carbon society, green cities and sustainable building as well as renewable energy and climate change. Other areas of expertise include capacity building and coaching training on strategic planning and leadership, developing vision, mission, and strategic goals, as well as action plans. Between December 2014 and till date, Dr. Aboulnaga has been appointed by the European Union and

Ministry of Foreign Affairs (MoFAs) as a Senior Adviser and National Expert of the "Clean Energy Savings Mediterranean Cities – CES-MED" project, a funded project by the European Union (2013–2018). In October 2015, Dr. Aboulnaga has been appointed as Senior Expert by Human Dynamics Consortium KG to develop the CES-MED National Recommendation Report (NRR) on Sustainable Urban and Energy Savings Actions (SUESA) for Egypt, as part of CES-MED project. In May 2016, he has been appointed by the EU and MoFAs as the National Expert of CES-MED project in Egypt – Associated to Transitions – Sustainability Consultant, Paris, France, to develop the Sustainable Energy Savings Climate Action Plan (SECAP) for the Governorate of Red Sea and Governorate of Luxor, Egypt. Dr. Aboulnaga has been Invited Expert and participated in the fifth Steering Committee Meeting (May 5), as part of the framework of the European Union-funded projects 'Cleaner and Energy Saving Mediterranean Cities' (CES-MED) and 'Sustainable Urban Demonstration Energy Projects' (SUDEP), and the high-level conference on the Role of Local Authorities in Sustainable Energy, Skhirat, Morocco, May 4–7, 2015. Furthermore, Dr. Aboulnaga is a Senior Expert on Sustainable Urban Development and Green Cities, Union for the Mediterranean (UfM), in Barcelona. Dr. Aboulnaga is a Former Strategy and Policy Advisor – Environment and Infrastructure – at the Prime Minister's Office of the UAE (February 2009–April 2010), responsible for implementing and monitoring the Federal Government Strategy (cycle 2008–2010). Prior to that, he held a position in the Capacity of Strategy and Policy Advisor at the Executive Council, Government of Dubai (May 2007–January 2009), where he developed many strategic projects, namely, Dubai Government Green Building Policy 2007, Dubai Heat Island Effect and Orthophoto Flyover Project in collaboration with LBNL in the United States, Detailed Policy for a Sustainable Dubai Urban Development Framework 2020 and Green Building Guidelines for UAE Ministry of Public Works and LAS (January 2009). Dr. Aboulnaga founded the Emirates Green Building Council in 2006 and was elected the first Chairman of the Council and Board of Directors. Dr. Aboulnaga is an Honorary Fellow of the Institute of Green Professionals, USA. He is also an Associate Partner at European Sustainable Development Network (ESDN) in

Austria since 2010 and a Registered Expert with UNEP and ESCWA since 1997. In December 2013, he participated in the European Commission Joint Research Centre (EC-JRC) expert meeting in Ispra, Italy, to review the Sustainable Energy Guidebook for ENPI South countries. Professor Aboulnaga is an author of 7 book chapters by Elsevier and Springer Nature; has been invited as keynote speaker and speaker to more than 380 international and regional conferences, seminars, workshops and training programmes; and has more than 165 international publications and presentations. In addition, he is an External Examiner of the Ph.D. panel at the University of Sheffield, UK. He taught at the University of Leeds, the American University in Dubai and the University of Dubai. Dr. Aboulnaga is currently teaching at Cairo University.

Amr F. Elwan is a Committed Lecturer with over 23 years of experience at leading Egypt and UK academic institutions, teaching architecture and environment for undergraduate and postgraduate students from various social and cultural backgrounds. He possesses an excellent administrative, verbal communication and written skills along with constructive and effective teaching methods that promote a stimulating learning environment.

He has many years of consultancy experience in sustainability, green urban development and sustainable built environment, including energy efficiency, green building, and green cities in developing countries, as well as academic positions. He worked as a LEED Consultant in Zweil Science City, where he held the position Head of Architectural Department at the Consulting Office, Engineering Authority of the Armed Forces. Dr. Elwan worked as Research Associate in the modules of interactive built environment at the University of Sheffield, UK, from 2010 to 2013. Also, Dr. Elwan holds a Lecturer position at the Arab Academy for Science, Technology, and Maritime Transport (AASTMT) teaching in the postgraduate programme and, in addition, is a Lecturer in the Department of Architecture at the Military Technical College (MTC) where he is teaching in the undergraduate and postgraduate programmes.

Dr. Elwan is Head of Architectural section at the Consulting office in El Afifi, Expertise House of Engineering Consultant, and Head of Architectural Department at the Consulting Office in Engineering Authority of the Armed Forces. He was also Head of technical office at National General Contracting and Supply Company - National Service, Armed Forces; and Head of the architectural section of the Consulting Office at the Military Technical College (MTC). He is currently a Lecturer in MTC till present. He is participating in the design process of most of the national projects in Egypt, namely, the 1.5 million acres with Ministry of Agriculture and Ministry of Irrigation and Water Resources. In addition, he developed full strategic planning and detailed plans for West Al Minia Algidida, and revised the drawings of New Capital that is coordinated by the Ministry of Housing, Utilities, and Urban Development (MoHUUD), the Ministry of Health (MoH), and 5+UDC Urban Development Consortium, as well as the "One million residential units" project and revised all governorates' administrative building borders with the Ministry of Local Development (MoLD) and finally, the design of a 200-bed hospital project in Badr City for the Egyptian Syndicate of Engineers. He also worked at the Ministry of Defense (MoD) within the team responsible for handling hundreds of projects along 20 years including: hospitals, office buildings, residential settlements, hotels, etc. Moreover, he worked with different consulting offices in the design of the new Headquarters of MoD and new Headquarters of Ministry of Interior (MoI) in the New Administrative Capital, Cairo, Egypt.

Dr. Elwan supervised many master's degree theses, and he has more than 12 international publications. He also attended many international conferences in Sweden and the UK. He has participated in a research team with Prof. Mohsen Aboulnaga (Team Leader) in an applied research project that was funded by AASTMT between 2015 and 2017 to upgrade a farm residential house to be sustainable and green building in al-Zarka village in Beheira Governorate, Egypt.

Mohamed R. Elsharouny has 8 years of experience in higher education and consultancy in project management and in energy efficiency, green building and climate change as well as academic positions. He is currently an Assistant Lecturer at the Faculty of Engineering, Modern Academy, Egypt (2016 till date). He is also Outsourced Senior Engineer at Afreximbank, where he is responsible for retrofitting of the Bank Headquarter in the area of energy efficiency, sustainability of spaces and long-term requirements and innovative, as well as cost-effective, operation, maintenance systems and processes.

Architect Elsharouny is competent Architect and Consultant with many skills in energy simulation programmes. His area of interests are sustainability, sustainable development and sustainable built environment, including energy efficiency and green building and cities in developing countries and worldwide. His areas of competency include climate change vulnerability assessment, climate adaptation measures and policies, low-carbon society, as well as renewable energy and water management. Other area of expertise includes building retrofitting and project management.

Between 2015 and till date, Mohamed is a Volunteer in many activities in supporting sustainable development. He is a Project Leader for Wiki Loves Earth and Wiki Loves Africa in Egypt that took place in 2017, which is a contest and workshops for heritage.

Mohamed is an author of a book chapter by Springer, which is currently under process in the press, and a speaker at more than five intentional conferences and has more than six international published papers and presentations.

Architect Elsharouny obtained his Bachelor of Science (B.Sc.) in Architectural Engineering from Modern Academy (2012) and Master of Science (M.Sc.) in Environmental Design from Cairo University (2018).

List of Boxes

List of Figures

List of Tables

Chapter 1
Introduction

Abstract Climate change is inevitable as it is manifested worldwide, and poses a fundamental and unprecedented threat to all sectors, regions and societies. This chapter highlights the background on climate change, and discusses the post COP 21 and COP 22 climate actions. It also highlights the size of the problem of greenhouse gas emissions and climate change risks. In addition, this chapter addresses the importance of sustainable development implementation in cities as a path to adapt to climate change. Furthermore, it presents the structure of the book, which is divided into two parts and encompasses eight chapters (including this chapter). Finally, a note from the lead author is presented.

1.1 Background

Climate change is inevitable! Climate change is manifested worldwide, and it poses a fundamental and unprecedented threat to all sectors, regions and societies. In a speech at one of the world's major gatherings on climate change, H.E. Hu Jintao, President of China from 2003 to 2013, made a strong statement and stressed that climate change has absolutely been manifested and seen in many parts of the world in the past decade [1]:

> Global Climate Change has a Profound Impact on the Survival and Development of Mankind. It is a Major Challenge facing all Countries.

It is clearly noticed that in the past 10 years climate change has hit different parts of the world from north to south and from west to east with no differentiation of boarders; whether developing or developed countries.The latter is known as high emitters' nations of greenhouse gas compare to low emitting nations. This has been truly said by the Chancellor of Germany, H. E. Angela Merkel [2]:

> Climate change knows no borders. It will not stop before the Pacific islands and the whole of the international community here has to shoulder a responsibility to bring about a Sustainable Development.

Recently, climate change has impacted many regions through severe events from the East Coast of the United States mainly North and South Carolina in September 2018

M. M. Aboulnaga et al., *Urban Climate Change Adaptation in Developing Countries*, https://doi.org/10.1007/978-3-030-05405-2_1

and the typhoons that destructively hit Japan twice in August and September [3] as well as Mangkhut storm that severely hit Hong Kong, China and the Philippines in September [4]. All caused massive destruction to the cities' infrastructure including roads, bridges and airports as well as cost the economy a hefty amount of money that affects the economy, besides left dead citizens and thousands of injuries. In addition, heat waves, which are considered climate change severe events, are continuously hitting many parts of the world. These are manifested in Japan and Europe, e.g., July to August 2018 and Egypt in August 2017.

Globally, negative consequences of climate change are represented in many parameters; sea-level rise, water scarcity, food insecurity, pressure on human health, ecosystems and the national economy. All climate risks, whether directly or indirectly, affect cities and infrastructure, which are already suffering from induced human pressure and absence of sustainable development. Climate change has enormous serious impacts which threaten the built environment, including lives, buildings and infrastructure, as well as hinder development around the world.

According to Global Humanitarian Forum, 500 million people are at extreme risk due to floods, droughts, landslides and tropical cyclones [5]. In this line, Ban Ki-Moon, former Secretary General of the United Nations, stated that climate change is annihilating the path to sustainable development [6]:

> Climate change is destroying our path to Sustainability. Ours is a world of looming challenges and increasingly limited resources. Sustainable development offers the best chance to adjust our course.

Based on future estimate models, the cost of climate change risks amounts to 2–10% or more of the global GDP per year, depending on the mean global temperature rise [7]. Nevertheless, peer-reviewed calculates of the social cost of carbon (net economic costs of damages from climate change aggregated across the globe and discounted to the present) for 2005 have an average value of US$12 per ton of CO_2, but the range from 100 estimates is large (−$3 to $95/t$CO_2$). According to IPCC, such breadth of published proof implies that the net damage costs of CC are projected to be noteworthy and to augment over time [8].

1.2 Post COP 22 and COP 21

> Our climate is warming at an alarming and unprecedented rate. We have an urgent duty to respond.

The COP 22, held between November 8 and 18, 2016, in Marrakech, Morocco, had ended with two main conclusions: (a) era of implementation and (b) commitment and actions. These actions are directed primarily to actions on climate and attaining sustainable development [9]. Therefore, the need for urgent response to climate change impacts is momentarily taking place to mitigate such impacts of severe events on the most vulnerable sectors, whether on the current risks or the future risks. Nonetheless, these responses are not at the same momentum with the severity of climate change risks. The Marrakech Action Proclamation for Climate and

Fig. 1.1 Paris Agreement on climate change – COP 21, December 2015. (Image Source: http://www.theguardian.com/environment/2015/dec/13/world-leaders-hail-paris-climate-deal)

Sustainable Development has been officially launched for the first time, where governments, world's leading institutions and civil society representatives had come together during COP 22. The good news is the Paris Agreement, signed on December 15, 2015, where at the Climate Change Conference culminated in a big success with the adoption of a new global climate change agreement, which is backed by 196 countries. This agreement is to mitigate climate change risks and yielded a global agreement to keep the earth's temperature at 1.5 °C – COP 21 [10] (Fig. 1.1).

Soon after the Paris Climate Change Agreement, China and the United States – the world's largest emitters of GHG (about more than 40% of the total world emission) – had signed the agreement during the G20 Meeting held in Hangzhou, China, September 2016 [11]. The good side of the coin is that the United States and China ratified the Paris Agreement (Fig. 1.2), but the bad side of the coin is that President Trump has officially pulled out of the Paris Agreement in April 2017. This action shocked the world and caused a dilemma about the 20 per cent of GHG emission generated by the United States.

With the current actions, it is imperative that more global actions are urgently needed to achieve the Paris Agreement targets; hence it is the role of all governments signed on the agreements to meet such targets. For the world to start meeting the Paris Agreement set goals, the International Renewable Energy Agency (IRENA) indicates that renewable energy requires to be expanded at least six times quicker [12]. At the moment, statistics of global emission trends are not on track to attain the goal. Government plans still fall far short of emission reduction needs. According to IRENA, with existing and planned policies, the world would exhaust its energy-related 'carbon budget' (CO_2) in under 20 years to keep the global temperature rise to well below 2 °C (with 66% probability), while fossil fuels such as oil, natural gas and coal would continue to dominate

Fig. 1.2 G20 Meeting in Hangzhou, China, September 2016. (Image Source: https://en.wikipedia.org/wiki/2016_G20_Hangzhou_summit#/media/File:G20_2016_leaders.jpg)

the global energy mix for decades to come [12]. It is vital to note that immediate action will be crucial to achieve further reduction of cumulative emissions by 470 gigatonnes by 2050 compared to current and planned policies (business as usual) in order to reach the set goal – to keep temperature below 2 °C [12].

Climate change adaptation measures reduce the adverse effects of climate change risks by taking appropriate actions to prevent or minimize the damage resulting in exposure to hazards or taking advantage of opportunities that increase resilience to climate change and sustainability. Also, climate change adaptation policies are seamlessly crucial in the efforts to lessen climate change risks. Additionally, climate change adaptation through projects and action plans are essential, particularly in developing countries, since these countries are most vulnerable to climate change severe events and associated impacts. Greenhouse gas (GHG) reduction is the way forward to reduce the impact of climate change. The International Renewable Energy Agency (IRENA) report 2018 indicated that sustainable energy (renewable energy and energy efficiency) could, if combined, offer over 90% of required energy-related CO_2 emission reductions [13].

The Center for Climate Strategies (CCS) developed and pursued an analysis and implemented a Low Emissions Development Strategies (GLEDS) programme in Guatemala (a developing county). This project was coordinated and funded by the US Agency for International Development (USAID) to meet the nation's NDC commitment and sustainable economic and social development goals [14]. Figure 1.3 presents a detailed design of 43 policy actions to meet national economic, energy, resource, emission and equity goals selected by a broad range of stakeholders and government agencies through an inclusive, consensus-based process, and Fig. 1.4 depicts the results of direct impact analyses for the 43 policies that cover GHG emission reduction, energy and resource shifts and indicates that if all policies were fully implemented, the expected 2050 GHG emissions could be reduced to levels seen prior to 2010.

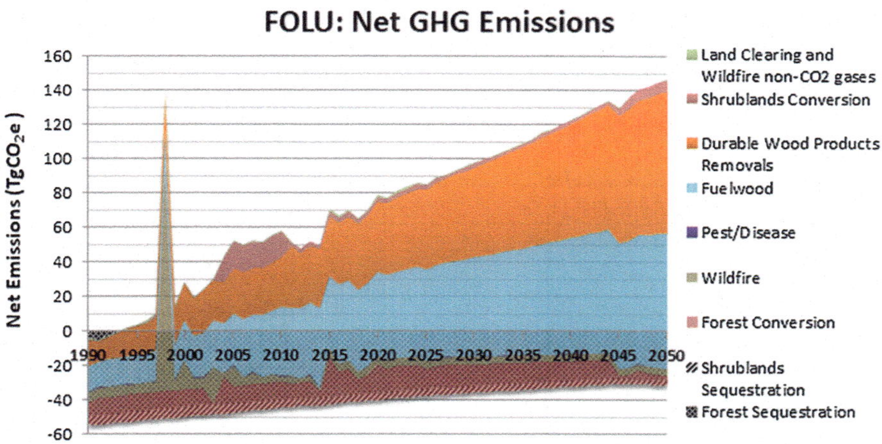

Fig. 1.3 Net greenhouse gas emissions resulting from policy actions to meet national economic goals. (Source: CCS - http://www.climatestrategies.us/international_actions/international_actions/view/8)

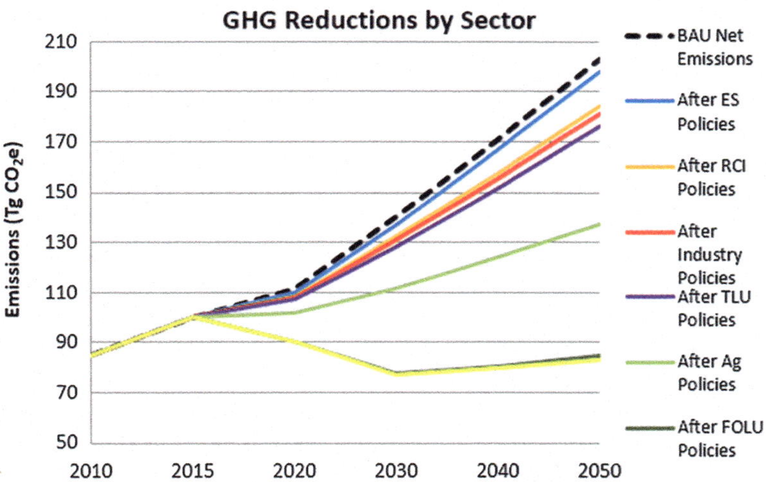

Fig. 1.4 Reduction in greenhouse gas emissions by sector resulting from direct impact analyses of policies. (Source: CCS - http://www.climatestrategies.us/international_actions/international_actions/view/8)

1.3 Climate Change: What Is the Size of the Problem?

By looking at the carbon stock in the ocean, atmosphere and the human-generated emissions per year and per person in developed countries versus developing countries [15, 16], it is clearly noticed that the carbon stock in the ocean is huge (140,000 billon tCO_2eq) compared to that in the atmosphere (2200 billon tCO_2eq), the latter is only 1.6 per cent of that of the ocean, whereas the human-generated emissions per year in Africa (6.97 tCO_2eq/year) is very low compared to the highest six emitting countries (61.96 tCO_2eq/year) as shown below.

Carbon stock in the ocean	140,000 billon tCO_2eq [16]
Carbon stock in the atmosphere	2200 billon tCO_2eq [16]
World human GHG emissions	53 billion tCO_2eq/year [16]
World human GHG emissions per capita	7.55 tCO_2eq/year [16]
GHG emissions – Egypt (2012)	295 million tCO_2eq/year [16]
GHG emission per capita – United States	16.22 tCO_2eq/year [15]
GHG emission per capita – China	7.45 tCO_2eq/year [15]
GHG emission per capita – India	1.92 tCO_2eq/year [15]
GHG emission per capita – Germany	9.47 tCO_2eq/year [15]
GHG emission per capita – Japan	9.68 tCO_2eq/year [15]
GHG emission per capita – Australia	17.22 tCO_2eq/year [15]
GHG emission per capita – Africa	6.97 tCO_2eq/year [15]
GHG emission per capita – Lebanon	3.64 tCO_2eq/year [15]
GHG emissions per capita – Algeria	3.85 tCO_2eq/year [15]
GHG emissions per capita – Egypt (2012)	3.44 tCO_2eq/year [16]
GHG emissions per capita – Egypt (2016)	2.29 tCO_2eq/year [15]

An example of the magnitude of climate change risks – in few days of the year 2017 in one city (Texas, USA) – was recorded in terms of damage and cost of destruction as follows: a) 20 million citizens in danger; b) 40,000 houses without power for 2 weeks; c) 100,000 houses that were damaged; and d) 11 billion cost of destruction in the city [17]. These records are shocking figures to the economy of the city and the USA.

In 2018, Japan has experienced an extremely deadly heat wave with a displayed temperature reaching 41.1 °C, the highest ever logged in Japan. This deadly heat wave has left 77 people dead and sent more than 30,000 to hospitals across the nation from July 9–22, 2018 [18].

Figure 1.5a depicts the soaring temperature, whereas Fig. 1.5b, c shows citizens under severe heat sipping water in Tokyo at a record high of air temperature reaching 41 °C and cooling down by playing in the water jets at a park near Nerima in Tokyo, Japan, respectively [19].

Europe was not far from the severe heat wave in 2018. In the United Kingdom, temperature soared at a recorded high on July 23, 2018; the hottest day so far recorded was 33.3 °C at Santon, Downham, in Suffolk. In result, citizens escaped the city's severe heat in Bournemouth, England, UK, and headed to beaches (Fig. 1.6b), whereas Britain is basking in record temperature, a record high of air temperature in Wales of 35 °C [20]. Also, cracks appeared in the dry banks of March Haigh reservoir, UK, which can be depicted in Fig. 1.6a. In addition, a severe heat wave struck Paris and northern France (Fig. 1.6c), followed by an official heat wave warning on Wednesday (July 15 and 23, 2018) due to high temperatures, which soared around the country (32–35 °C).

(a) (b) (c)

Fig. 1.5 Japan deadly heat wave of July 2018. (Image sources: (**a**) https://www.express.co.uk/news/weather/993592/.(**b**)https://asia.nikkei.com/Life-Arts/Life/Japan-logs-over-41-C-for-new-record-high-as-deadly-heat-wave-continues. (**c**) https://www.bbc.com/news/world-asia-44921834)

(a) (b) (c)

Fig. 1.6 (**a**–**c**) The United Kingdom and France's extreme heat waves of July 2018. (Image sources: **a**. https://www.express.co.uk/news/weather/984996/. **b**. https://www.euractiv.com/section/climate-environment/news/scientists-see-clear-evidence-of-climate-change-behind-2017-heatwave/. **c**. https://www.thelocal.fr/20180724/heatwaves-paris-placed-on-alert-as-mercury-rises)

In Germany, heat waves also hit many parts of the country during July 2018. The heat wave reached up to 36 °C and temporarily shut downed Hannover Airport (41 departures and 44 arrivals were affected by major delay) due to cracks on the runway, as shown in Fig. 1.7a. In addition, high temperatures resulted in many areas of crops devastated (Fig. 1.7b), which caused a loss of US$1.6 billion. Furthermore, the drought and heat wave led to the drying out of the riverbank of the Elbe in Magdeburg [21].

In the MENA region, the World Bank stated the fact, based on the IPCC computer climate modelling project results, it is predicted that sea levels rising by over 0.5 m by the end of the century would place low-lying coastal areas in Tunisia, Libya, UAE, Kuwait and Egypt at particular risk [22]. Hence, climate change adaptation including strategy, policies, measure and actions is crucial to address such impacts.

Throughout history, civilizations have been associated with climate. In some parts of the world, people and societies have adjusted and coped with climate at varying degrees of success. Drought, for example, has been at least partly responsible for the

(a) (b)

Fig. 1.7 Extreme heat wave damages – Germany, July 2018. (**a**) Cracks damage the runway. (**b**) Large areas of devastated crops. (Image source: **a.** http://www.haz.de/var/storage/images/haz/hannover/aus-der-stadt/sperrung-am-airport-flughafen-zieht-erste-konsequenzen/695272156-3-ger-DE/Hitzedefekt-Flughafen-Hannover-zieht-erste-Konsequenzen_reference_2_1.jpg. **b.** https://phys.org/news/2018-07-drought-germany-breadbasket.html)

rise and fall of civilizations. Scientifically, the idea of adaptation emerged in 1991 by the US National Academy of Sciences [23].

1.4 Why Sustainable Development in Cities Is a Path to Climate Change Adaptation?

In fact, cities around the globe are encountering various strategic challenges to meet green criteria, standards and sustainable development measures. But nevertheless, in many countries, the 'business-as-usual' (BAU) cities' models which are still operated lead to major defies. The International Resource Panel (IRP) report issued by the United Nations Environment Programme (UNEP) indicated that cities have a vital role when it comes to offsetting climate change impacts [24].

Cities produce 80 % of the global GDP on just 2 % of the Earth's land surface.

In addition, cities currently consume 60–80% of the global total energy (electricity, transport, etc.) and are responsible for 75% of the natural resources and 70% of carbon dioxide (CO_2) emissions [24]. Asia is already facing vast environmental challenges, as 11 of the 20 most polluted cities in the world and 3 of the top 5 CO_2-emitting economies are in Asia.

Hence, cities and their infrastructure should be designed in ways that they are less resource, emission intensive, which create a less polluted, healthier environment for their residents and to assist in the efforts towards climate change adaptation. Another challenge to attain sustainable development and adapting to climate change impacts is the fact of increasing population. The OECD predicted that with the estimated 9 billion people, the economy would need to be 15 times the size of today's economy (75 times bigger than what it was in 1950) and 40 times bigger than today's economy – 200 times bigger than in 1950 [25].

It is important to understand how this affects adaption measures and actions. This is a major defy to be addressed in the efforts of climate change mitigation and adaptation. Furthermore, in developing countries, the story is different.

1.5 Our Converge

The prime focus of this book, in its broad sphere, is *climate change* and *built environment* with particular thought in mind on climate adaptation of cities and its infrastructures since climate change impacts human settlements and is manifesting worldwide.

One of the challenges drawn from the literature is that Africa is one of the main continents that are vulnerable to climate change risks due to many factors. This book is addressing an important topic that attracts the world attention at the moment – urban climate change adaptation in developing countries. The book is structured into two parts that encompass in total seven chapters. Part one focuses on climate change impacts on urban areas and infrastructure as well as climate change risks and sectors' and variables' changes. Part two addresses global climate change adaptation measures in cities and urban areas and policies; and presents risks, policies, measures, and actions for climate change adaptation in developing countries with emphasis on adaptation projects. Finally, it presents simulation and its results of climate change adaptation scenarios in 2080 in selected urban areas in developing countries.

1.5.1 Part I

This volume is dedicated to presenting some of the key issues of climate change risks around the globe, mainly in urban areas and infrastructure. This part is outlined into two chapters (Chaps. 2 and 3).

Chapter 2 is devoted to climate change as a major threat to the built environment, where the impacts of climate change affect natural and human systems all over the world. This chapter also highlights the impacts of climate change on activities and sectors in order to investigate their consequences on the built environment. Global climate change impacts on different sectors such as agricultural and food security, ecosystems, forests, water, health, coastal and flood-prone zones, tourism and energy have been discussed. Such impacts are summarized and analysed based on three factors: effect (positive or negative impact), occurrence (impact occurred or will occur in the future) and region (continents affected by impact). In addition, it explains the risks and rewards of climate change, along with the sectors that are mostly affected. Moreover, it determines the relative weight of the negative or positive impacts of climate change. Furthermore, the status of impacts in terms of occurrence was also listed, and continents were ranked according to their degree of exposure to climate change risks to assist in forming better assessment for potential impacts of global climate change.

Chapter 3 focuses on climate change impacts on human settlements that are manifested worldwide. Such impacts arise from climate change risks, notably sea-level rise (SLR), desertification, drought, extreme events, food insecurity, increased health risks and temperature-related morbidity in urban environments. Existing trends towards urbanization has been witnessed due to the negative impacts of climate change on rural livelihoods. Also, settlement patterns' changes may not be restricted to the socially vulnerable, but will be extended to coastal areas' dwellers and SLR, which will force people to migrate to unaffected areas. Large population movements caused by deteriorated environmental conditions may lead to conflict through competition for resources in the receiving area. This chapter investigates the direct and indirect impacts of climate change on urban areas and infrastructure. The indirect impacts are reflected on the built environment and consequently on the economy of urban and rural regions. Moreover, the chapter explores and argues whether all climate change risks occurred or not. Finally, it determines which continents that are most exposed to climate change risks related to the built environment. Thus, it addresses the direct and indirect impacts of climate change on the built environment with a wider view on its cost.

1.5.2 Part II

This volume encompasses five chapters dedicated to presenting some of the key issues of climate change measures globally. It also depicts climate change adaptation policies and discusses its risks. Large portion of this part is dedicated to climate adaptation policies, measures and actions for adaptation in developing countries with focus on climate change adaptation projects and adaptation scenarios by 2080 in urban areas in developing countries.

Chapter 4 identifies adaptation measures in cities, including urban and rural areas that aid in preventing, moderating or adapting to the consequences of CC risks. These measures are discussed specifically on ten sectors such as flood-prone areas, urban areas, infrastructure and activities in the city such as fresh water, energy, public health, agriculture, forest, biodiversity and economy. Also, the share of Africa, Asia, Europe, North America and Latin America from these measures will be identified. Finally, the chapter argues and explains the role and significance of CCA measures in upgrading cities and enhancing buildings' performance.

Chapter 5 focuses on climate change policies. It consists of three folds: the first fold investigates climate change adaptation (CCA) policies that were adopted in Africa, Asia, Europe, North America and Latin America. These CCA policies are related to urban areas, infrastructure, water resources and energy as well as green infrastructure, public health and tourism sectors. These adaption policies have a significant role in cities' upgrading and buildings' retrofitting. The second fold identifies the mechanism of forming and prioritizing CCA policies to be integrated into the climate action plans. The third fold determines the challenges and barriers appeared during adaptation process; thus it suggests the method to overcome these obstacles for better adaptation practices. The importance of this chapter stems from discussing CCA policies that are responsible for managing CC measures in a large scale, whether on a city level or a country level to avoid conflict with other measures and developing plans and facilities that can be integrated into existing policies. Nonetheless, by adopting policies only, this doesn't seem useful; thus discussing tools such as adaptation decision framework is vital to shove adaptation from theory phase to practice and yet learn how to overcome the obstacles it faces for successful climate change adaptation.

Chapter 6 investigates climate change adaptation in developing countries. In line with the fact that Africa is the most vulnerable continent, therefore Egypt, as one of the developing countries, is selected as a case study since it is one of the most exposed countries to climate change risks. Climate change impacts, adaptation policies and adaptation measures in developing countries were reviewed, presented and discussed. The chapter is laid out in three main folds: the first part discusses climate change risks on urban and coastal areas, infrastructure and related economic sectors including agriculture, public health, ecosystem, fishing, tourism and politics; and the second and third folds highlight the adaptation policies and measures adopted by government authorities and civil society. These policies, strategies and measures include building up the scientific capacity, developing adaptation strategy to increase energy and water efficiency, enhancing land use management, improving public health and retrofitting vulnerable buildings and infrastructure. It is important to note that this chapter examines how the action plans address the risk of climate change in order to cope with the impacts or consequences of climate change in the short or medium term.

Chapter 7 presents, examines and argues climate change adaptation projects in developing countries by comparing nine projects in one of the developing countries – Egypt – and global case studies in other developing countries to identify the gaps and track success stories and failures to achieve adaptive learning. In the comparative analysis, nine adaptation projects in Egypt were selected and emphasized on six sectors: energy, urban areas and awareness raising as well as waste management, ecosystem and agriculture. These sectors were compared to similar projects that were implemented as part of the climate action plans. These projects represent Morocco, Mexico, Brazil, Kiribati, India, the Philippines, Solomon Islands, Central African Republic and Mauritius. The chapter starts with an introduction about climate change adaptation projects and clarifies the scope of study and then discusses and validates the evaluation criteria that depend on seven main features, relevance, effectiveness, sustainability, equity, efficiency, flexibility and stakeholders' participation, and converts them from qualitative to quantitative in order to be measured. The evaluation of the climate change adaptation projects includes an overview about each project and identifies the project's description, scope, timeline, aim, challenge and outcome which are required for such evaluation. The results of the comparison were analysed and the gaps were identified.

Chapter 8 presents the assessment of climate change (CC) impacts and scenarios as well as the simulation measures and tests their applicability in urban areas in order to anticipate and predict climate change impacts and scenarios. It also highlights the selected case study that is illustrating a housing project in an urban area in a developing country, which is characterized as a hot-arid climate. In addition, the study simulates the urban microclimate in terms of the urban metaphor and setting from 2016 (baseline year) and compares it to the scenarios in 2080 in order to evaluate CC impact. Furthermore, it tests and analyses six measures including (a) cooling passive design configurations, (b) buildings' elevation, (c) buildings' envelope, (d) vegetation, (e) water features and (f) orientation and high albedo, to identify their impact on the adaptation scenarios in terms of heat stress and comfort. The climate change weather scenarios for 2080 were generated exploiting CCWorldWeatherGen, which is based on the IPCC Third Assessment Report. Moreover, the chapter scrutinizes the role of environmental design in climate change adaptation (CCA) in cities since sustainable urban development is a part of climate action. Thus, it is imperative to investigate if such action was implemented on new projects and weather forecast meets the climate change adaptation measures. Finally, the climate change adaptation scenario 2080 results were presented and discussed.

1.6 A Final Note

Why do we urgently need *green cities* and to *adapt to climate change*? In fact, cities and their infrastructure should be designed in ways that they are less resource and emission intensive and which create a less polluted, healthier environment for their residents.

The OECD published on September 25, 2018, a report highlighting 'Financing Climate Futures: Rethinking Infrastructure', in which scaling up and shifting financial flows to low-emission and resilient infrastructure investments are critical to delivering on the objectives of the Paris Climate Change Agreement and the 2030 Agenda for Sustainable Development and SDGs [26].

Hence, it is imperative to note that OECD climate change work is focusing on how to shift countries to a low-carbon and climate-resilient pathway yet how to enhance the effectiveness of the global climate regime. Building and upgrading cities to be green and resilient would contribute to this notion and significantly reduce the impacts and risks of climate change.

There is an urgent need for developing green and sustainable cities, and upgrading conventional cities to adopt and incorporate features of a green city such as renewable energy, energy efficiency, and managing material consumption as well as waste management and recycling – meaning green building and increasing green coverage including green roofs (Figs. 1.8, 1.9, and 1.10).

Green building and green roofs form large parts of green cities.

Fig. 1.8 Green coverage and green roof in EWA University, Seoul, South Korea. (Photo credit: Lead author)

Fig. 1.9 Egypt set renewable targets – 20% by 2022 and 40% by 2035. (Photo credit: Lead author)

> Sustainable energy (renewable energy and energy efficiency) is becoming financially and economically viable to support developing smart and green cities.

In this line, Gregory Robertson, Mayor of Vancouver, Canada, indicated that the cities must depict leadership to offset climate change risks and challenges by emphasizing on the role of renewable energy in a major and a full shift by using 100% clean energy [27].

Fig. 1.10 Solar PV panels for generating electrical energy from renewable sources. (Photo credit: Lead author)

The actual global energy consumption as of July 25, 2018, was reported, as total energy used, of the order of 306.4 TWh, of which 81.1% (248.4 TWh) is from non-renewable sources and only 19.2% (58.3 TWh) from renewables. Such energy consumption resulted in 104.6% increase from the 2017 CO_2 emissions (10.9 billion tCO_2) [28].

> Cities around the world must show continued leadership to meet the urgent challenge of climate change, and the most impactful change we can make is a shift toward 100 percent of our energy being derived from renewable sources.

There are colossal costs of inefficient energy use in the world and in developing countries. Unsustainable cities in the absence of Inclusive Green Economy strategies and policies would lead to more climate change severe impacts and manifestations in the coming years. Raffaello Cervigni, Regional Coordinator for Climate Change of the World Bank, has stated that 'although the MENA region contributes low emissions compared to global emissions (about 6 %), it is highly vulnerable to climate change impacts in strategic sectors, e.g., agriculture, water resource management and urban development' [29].

Developing cities infrastructure, specifically buildings and transport (60% of the city's main emitter sectors), should crucially address sustainable development measures and actions, as well as adhere to urban development policies. Thus, it is imperative to address, monitor, and report SDGs attainment, particularly the local government in developing countries to create sustainable cities.

Not far from the date of printing this book, climate change impacts have been witnessed in many cities worldwide, for example, in many parts of Mexico beach and Florida, USA (October 2018), where the strongest storm (Michael) since 1992 had left many parts of the city destructed and devastated and tore many buildings and city infrastructure (Fig. 1.11). Recently, severe climate change risks have been witnessed in Jordan (October 22 2018) as shown in Fig. 1.12. In October 2016, severe storms hit Upper Egypt and Red Sea Coast and damaged part of Ras Ghareb town and floods caused dis-colouring of the River Nile in Cairo (Fig. 1.13).

(a) (b)

Fig. 1.11 Climate change and Florida's destructive storm Michael, USA. (Source: (**a**) https://en.wikipedia.org/wiki/File:Mexico_Beach,_Florida,_after_Hurricane_Michael_2018.png (**b**) https://media.defense.gov/2018/Oct/13/2002051086/-1/-1/0/181011-A-PR094-242.JPG)

Fig. 1.12 Climate change impacts in Jordan, severe storms and floods (October 25, 2018). (**a**) Destructive floods in Mleeh and Madba. (Source: https://Jorda.times). (**b**) Severe floods near Dead sea. (Source: https://upload.wikimedia.org/wikipedia/commons/d/d2/%D8%B3%D9%8A%D9%88%D9%84_%D9%85%D9%84%D9%8A%D8%AD-%D9%85%D8%A7%D8%AF%D8%A8%D8%A7.jpg)

Fig. 1.13 Climate change impacts in Egypt, severe storm dis-coloured River Nile, Greater Cairo resulting from floods in Ras Ghareb, Red Sea coast (Oct 22, 2018). (**a**) and (**c**) The dis-colour of River Nile branch in Giza. (**b**) The dis-colour of River Nile branch in Cairo. (Source: Lead author)

A resilient and sustainable future requires making a leap forward towards climate change adaption and achieving sustainable development goals. Collective efforts are needed by all stakeholders to achieve the SDG targets and indicators by 2030 and lessen climate change risks and reduce economic losses.

References

1. "Hu Jintao's Speech on Climate Change", The New York Times, 22 September 2009. [Online]. Available at: https://www.nytimes.com/2009/09/23/world/asia/23hu.text.html. (Accessed: October 1, 2018).
2. Safi, M., "Angela Merkel pressures Australia to reveal its greenhouse gas targets", The Guardian, 17 November 2014. [Online]. Available at: https://www.theguardian.com/world/2014/nov/17/angela-merkel-pressures-australia-to-reveal-its-greenhouse-gas-targets. (Accessed: April 3, 2017).

3. McCurry, J., "Typhoon Jebi: Japan hit by strongest storm of 25 years", The Guardian, 4 September 2018. [Online]. Available at: https://www.theguardian.com/world/2018/sep/04/typhoon-jebi-japan-hit-by-strongest-storm-for-25-years. (Accessed: October 8, 2018).
4. "Typhoon Mangkhut battered the Philippines and coastal China, killing 69 and injuring dozens more", Business Insider, 17 September 2018. [Online]. Available at: https://www.businessinsider.com/typhoon-mangkhut-death-toll-china-philippines-hong-kong-2018-9. (Accessed: October 8, 2018).
5. "Understanding Risks (2012). Best practice in Disaster Risk Assessment, Proceedings from the UR 2012 Forum", The World Bank, [Online]. Available at: https://understandrisk.org/wp-content/uploads/ur2012_proceedings.pdf. (Accessed: October 1, 2018).
6. Ban Ki-moon Quotes, "Ban Ki-moon Quotes", [Online]. Available at: https://www.brainyquote.com/quotes/ban_kimoon_643779#IlvMZiTtFQzg09og.99. (Accessed: April 3, 2017).
7. Harris, J. M., Roach, B., and Codur, A., "The Economics of Global Climate Change, Global Development and Environment Institute", 2017. [Online]. Available at: http://ase.tufts.edu/gdae. (Accessed: October 1, 2018).
8. The Intergovernmental Panel on Climate Change (IPCC), "Climate Change 2007: Impacts, Adaptation and Vulnerability", Cambridge University Press, 2007.
9. "Marrakech Climate Change Conference - November 2016", [Online]. Available at: https://unfccc.int/marrakech-climate-change-conference-november-2016. (Accessed: October 1, 2018).
10. Vidal, J. et al, "World leaders hail Paris climate deal as 'major leap for mankind", The Guardian, 13 December 2015. [Online]. Available at: https://www.theguardian.com/environment/2015/dec/13. (Accessed: October 1, 2018).
11. Phillips, T., Harvey, F., and Yuhas, A., "Breakthrough as US and China agree to ratify Paris climate deal", The Guardian, 2016. [Online]. Available at: https://www.theguardian.com/environment/2016/sep/03/breakthrough-us-china-agree-ratify-paris-climate-change-deal. (Accessed: October 2, 2018).
12. "Global Energy Transformation: A Roadmap to 2050, Executive Summary", IRENA, 2018.
13. "Global Energy Transformation: A Roadmap to 2050", IRENA, 2018. [Online]. Available at: http://www.irena.org/publications/2018/Apr/Global-Energy-Transition-A-Roadmap-to-2050. (Accessed: October 1, 2018).
14. "GUATEMALA- CCS works with RTI International on USAID-funded Guatemala Low Emissions Development Strategies Project", Center for Climate Strategies, 2018. [Online]. Available at: http://www.climatestrategies.us/international_actions/international_actions/view/8. (Accessed: October 1, 2018).
15. "World Data Atlas", Knoema, [Online]. Available at: https://knoema.com/atlas/. (Accessed: October 1, 2018).
16. "Cleaner Energy Saving Mediterranean Cities CES-MED), SECAP document", The European Commission, 2017. [Online]. Available at: http://www.ces-med.eu/publications/seap-documents. (Accessed: October 5, 2018).
17. "Harvey floods in Texas", The Guardian, 2017. [Online]. Available at: https://www.theguardian.com/us-news/live/2017/aug/31/texas-flooding-chemical-plant-harvey-live/. (Accessed: October 1, 2018).
18. "JABI typhoon", Japan Times, 2018. [Online]. Available at: https://www.japantimes.co.jp/news/2018/07/23/national/temperature-tokyo-hits-nationwide-high-year-mercury-rises-40-8-western-city-ome/#.W8ICkHszZ6o/. (Accessed: July 19, 2018).
19. HOPPS, K., "Japan heat wave 2018: Japan declares NATURAL DISASTER - death toll mounts", Express, 2018. [Online]. Available at: https://www.express.co.uk/news/weather/993592/japan-heatwave-2018-natural-disaster-death-toll-tokyo-weather-forecast. (Accessed: October 3, 2018).
20. Simon, F., "Scientists see 'clear evidence' of climate change behind 2017 heat wave", Euractiv, 2017. [Online]. Available at: https://www.euractiv.com/section/climate-environment/news/scientists-see-clear-evidence-of-climate-change-behind-2017-heatwave/. (Accessed: October 5, 2018).

21. "Record drought grips Germany's breadbasket", Phys.Org, 2018. [Online]. Available at: https://phys.org/news/2018-07-drought-germany-breadbasket.html. (Accessed: September 24, 2018).
22. "Adaptation to Climate Change in the Middle East and North Africa Region", The World Bank, [Online]. Available at: http://web.worldbank.org/archive/website01418/WEB/0__C-152.HTM. (Accessed: October 1, 2018).
23. "Responding to Climate Change", NASA's Jet Propulsion Laboratory, [Online]. Available at: https://climate.nasa.gov/solutions/adaptation-mitigation/. (Accessed: October 1, 2018).
24. "International resource panel, 10 key messages on Climate Change", UNEP, [Online]. Available at: http://wedocs.unep.org/bitstream/handle/20.500.11822/21558/IRP_10key_messages_climate_change.pdf?sequence=1&isAllowed=y. (Accessed: September 24, 2018).
25. "OECD Environmental Outlook to 2050: The Consequences of Inaction - Key Facts and Figures", OECD, 2018. [Online]. Available at: http://www.oecd.org/env/indicators-modelling-outlooks/oecdenvironmentaloutlookto2050theconsequencesofinaction-keyfactsandfigures.htm. (Accessed: October 1, 2018).
26. "Climate change", OECD, 2018. [Online]. Available at: http://www.oecd.org/env/cc/. (Accessed: October 1, 2018).
27. Leahy, S., "Vancouver commits to run on 100% renewable energy", The Guardian, 2015. [Online]. Available at: https://www.theguardian.com/environment/2015/apr/10/vancouver-commits-to-run-on-100-renewable-energy. (Accessed: October 1, 2018).
28. "International Energy Outlook 2018", EIA, 2018. [Online]. Available at: https://www.eia.gov/outlooks/ieo/. (Accessed: October 1, 2018).
29. "Water Is Focus of Climate Change in Middle East and North Africa", The World Bank, [Online]. Available at: http://web.worldbank.org/archive/website01418/WEB/0__C-151.HTM. (Accessed: September 29, 2018).

Chapter 2
Global Climate Change Risks: Sectors and Variables' Changes

Abstract Climate change is a global phenomenon that refers to changes in weather patterns such as temperature, precipitation and wind which has an effect that last for an extended period of time. The impacts of climate change affect natural and human systems all over the world, and its risks can be observed in several sectors. This chapter highlights the impacts of climate change on activities and sectors in order to investigate their consequences on the built environment. The chapter also discusses global climate change impacts on different sectors such as agricultural and food security, ecosystems, forests, water, health, coastal and flood-prone zones, tourism and energy. Such impacts are summarized and analysed based on three factors: effect (positive or negative impact), occurrence (impact occurred or will occur in the future) and region (continents affected by impact). In addition, it explains the risks and rewards of climate change, along with the sectors that are mostly affected. Moreover, it determines the relative weight of the negative or positive impacts of climate change. Furthermore, the status of impacts in terms of occurrence was also listed, and continents were ranked according to their degree of exposure to climate change risks to assist in forming better assessment for potential impacts of global climate change.

2.1 Global Climate Change Impacts

Climate change (CC) is happening in the past decades and manifested in various parts of the globe. Climate change refers to major changes in weather patterns such as temperature, precipitation and wind, among other effects that last for an extended period of time over several decades or longer. This phenomenon results in more floods, droughts or heat waves happening that present challenges to buildings and cities [1]. Current trends indicate that CC happened due to an increase in greenhouse gases (GHGs) in the atmosphere in recent years resulted from burning large quantities of fossil fuels and deforestation. The first step in solving this problem is through diagnosis of climate change (CC) impacts; therefore this chapter summarises and identifies CC risks nine sectors. It also attempts to understand how

M. M. Aboulnaga et al., *Urban Climate Change Adaptation in Developing Countries*, https://doi.org/10.1007/978-3-030-05405-2_2

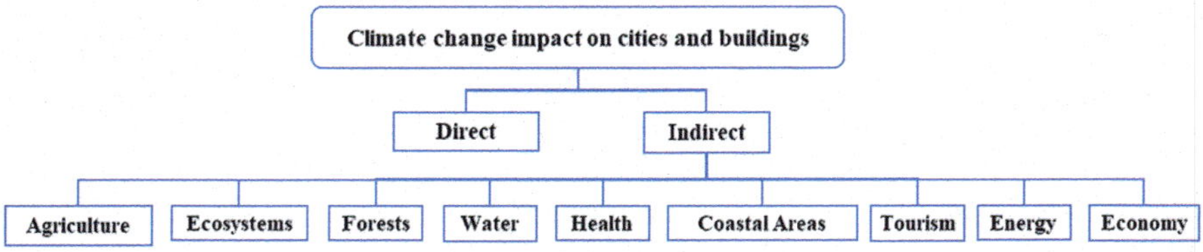

Fig. 2.1 Climate change impacts on cities and buildings. (Source: After Intergovernmental Panel on Climate Change)

much CC risks affect humanity, undermine global economy and threaten urban areas and infrastructure.

In recent decades, changes in climate have caused impacts on natural and human systems across all continents. Evidence of climate change impacts was observable in different sectors which also have consequences for social and human aspects of the built environment. These consequences on the built environment are divided into direct and indirect impacts; direct impacts affect cities following changes in climatic variability, while indirect effects result in consequences including environmental and social changes on sectors related to the built environment, which is known as economic of urban or rural areas.

Global climate change affects people and the environment in many ways. The potential effects include many different sectors, as shown in Fig. 2.1, such as agricultural and food security, ecosystems, forests, water, health, coastal and flood-prone zones as well as tourism, energy and economy. A selection of key findings regarding the impacts of climate change on systems, sectors and continents was done while determining which outweighs the other, assessing the negative or positive impacts of climate change and arranging the continents in the degree of exposure to climate change and then concluding their consequences on the built environment vulnerability.

The following section discusses climate change consequences on agriculture as one of most important economic sectors for rural areas.

2.1.1 Agriculture and Food Security

Agriculture is the most affected sector by climate change, as it depends on climate conditions. Thus, it's very sensitive to climatic variables changes. Negative impacts of CC on agricultural leads to an increased risk of starvation and diseases due to poor nutrition and high food prices that will cause harm, especially in developing countries [2]. According to IPCC, reduced crop productivity is the biggest risk on the international scale in the agriculture sector. Western Europe already suffers from yield losses in wheat due to increased heat stress [3] and reduced crop productivity due to fluctuations in average temperatures, rainfall and climate extremes, spread in pests and diseases and changes in atmospheric carbon dioxide (CO_2) and ground-level ozone concentrations and sea level rise (SLR) [4].

Soil erosion will threat Africa, Asia and Latin America [5]. Soil erosion is caused by water or wind: water by removing soil layers caused by the raindrop splash and runoff water or wind by sandblasting of seedlings or transplants. As a result of soil erosion, disease will increase, yield will decrease and quality and market value will be lost [6]. Desertification already threats Africa, Asia and Latin America [7]. Desertification means sand encroachment of vegetation which causes land loss, land degradation and difficulty in vegetation growth, especially to farmland adjacent to desert [8]. Desertification already affects one quarter of the total land surface of the globe as shown in Fig. 2.2. Desertification is caused by drought due to climatic variations such as high temperatures and irregular rainfall [9]. Increased pest and disease due to high temperatures in winter will be a suitable weather for mosquitoes to attack crops in spring which threats food security in Africa and Europe [10, 11].

Changes in crop suitability, growing season and the timing of cycle of agricultural crops in Europe increase the uncertainty about what to plant, where and when due to changes in precipitation, temperature and extreme weather events [12, 13]. Soil fertility and rain-fed crops' yield will be reduced in Africa due to the decrease in water quantity and quality [14, 15]. Growing seasons may lengthen in some areas in Africa and Latin America due to mild and increased rains [16, 17]. Crop productivity will increase in some regions in Europe and already increased in some regions in Asia. In the future, agriculture may expand in productivity in northern areas in Latin America and Europe [16]. However, there are three positive impacts of climate change such as increased in crop productivity, reduction of the growing period of crops due to increasing atmospheric CO_2 concentrations and indirect effects such as an increase crop growth and yield [18]. Most of the CC impacts on agriculture are negative (75%), and only 25% are positive (Fig. 2.3) which demonstrates the seriousness of the negative impact of climate change on agriculture.

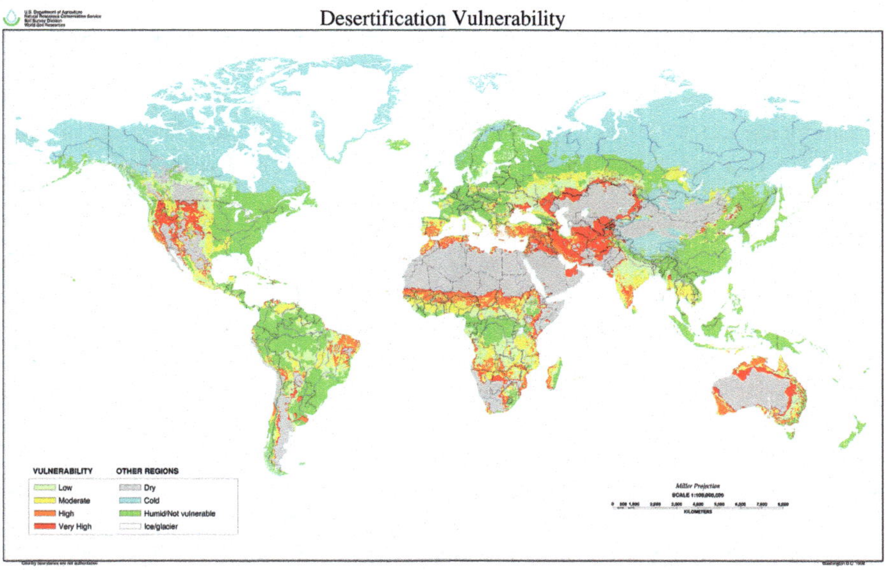

Fig. 2.2 Areas vulnerable to desertification. (Source: NRCS, 1998)

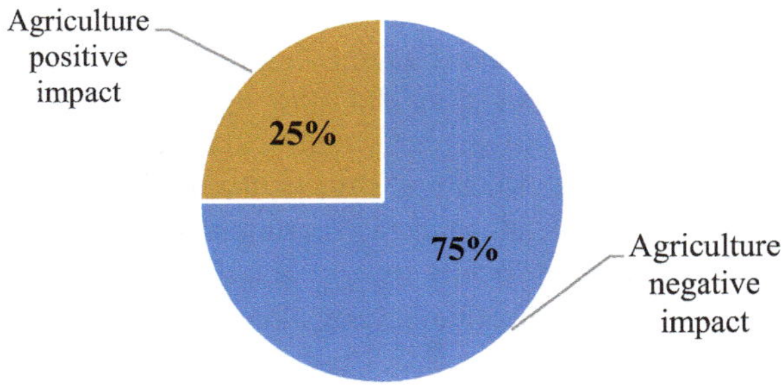

Fig. 2.3 Percentage of climate change impacts on agriculture. (Source: Developed by authors)

Box 2.1 Agriculture and Food Security
According to UNESCO, desertification already affects one-quarter of the total land surface of the globe, and roughly one third of the world's land surface is threatened by desertification due to climatic variations such as high temperatures, irregular rainfall and human activities as overgrazing and intensive farming [9].

2.1.2 *Ecosystems*

Loss of species and extinction of many plants and animals will be a threat worldwide, and such threat is already noted in Asia due to reduced food and water, increase of extreme weather events and wild fires [19]. The average across 131 model assumptions, the highest predicted extinction risks from CC, is in South America and New Zealand, which amounts to 14–23%, and the lowest risk is in North America and Europe (5–6%) [20]. Ice cap on mountains could disappear in Africa and Asia, and also glaciers will melt due to increases in carbon dioxide (CO_2) concentration. Global warming will also cause floods, reduce water availability and threat polar species as well as push them to shift north to higher elevations in Asia, North America and Latin America in the future to find tolerable thermal zone [21, 22]. Figure 2.4 shows the cumulative change in mass balance of glaciers worldwide; negative values indicate a net loss of ice and snow compared with 1945.

The frequency of rock falling will increase in North America and already increased in Europe due to rise in temperatures that melt glaciers which will destabilize mountains and threaten mountain valleys [23].

Spread of warm water species in Europe is another positive impact of CC. In some high-latitude regions, an increase of these species is due to global warming and decreased ice cover; however in low-latitude regions, some species may decline because fish cannot adapt to high temperatures and increased ocean acidity [24, 25]. Most of the CC impacts on ecosystems are negative 87% of the total impacts, and only 13% are positive (Fig. 2.5). Thus, this demonstrates the seriousness of the negative impact of climate change on ecosystems.

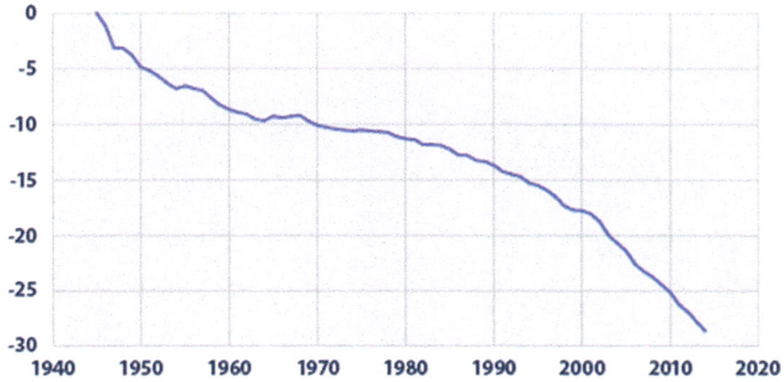

Fig. 2.4 Cumulative change in mass balance glaciers worldwide. (Source: US Environmental Protection Agency)

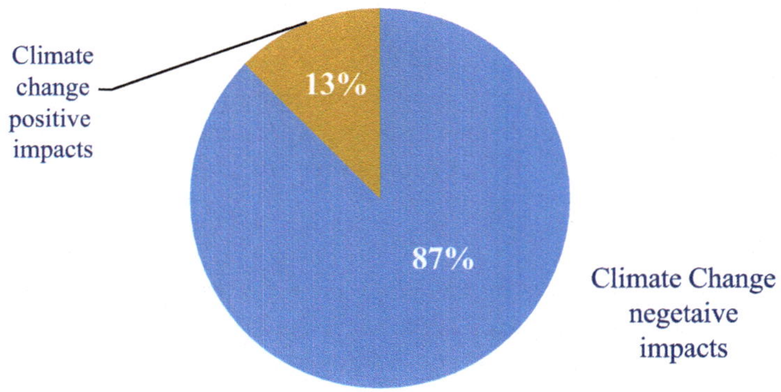

Fig. 2.5 Percentage of climate change impacts on ecosystems. (Source: Developed by authors)

Box 2.2 Ecosystems
Based on averaged across 131 model assumptions, the highest predicted extinction risks from CC in South America and New Zealand estimated 14–23%, and the lowest risks in North America and Europe estimated 5–6% [20].

2.1.3 *Forests*

According to the World Wide Fund for Nature (WWF), forests represent home for 300 million people and 80% of the terrestrial biodiversity worldwide, therefore, 1.6 billion people depend on them for their livelihoods [26]. Forest fires already threaten all continents (Fig. 2.6). Due to droughts and increased temperatures, it led to economic and lives losses and threat to animals and plants [27, 28]. Human activities

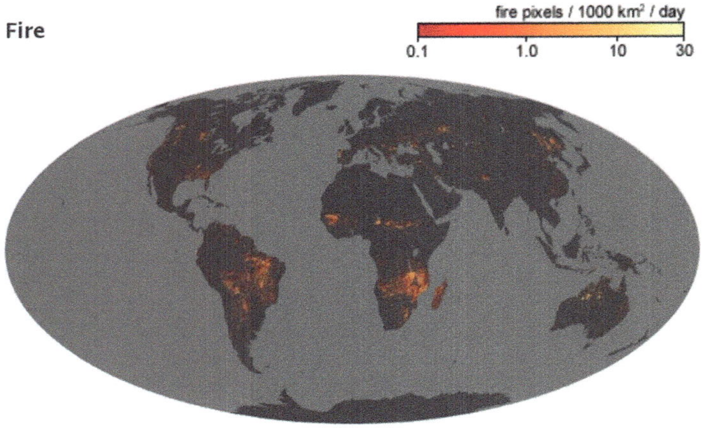

Fig. 2.6 Forests fire around the world. (Source: NASA)

also play a role in accelerating forest fires, for example, burning the forests in Indonesia to exploit as agricultural land [29]. Northward shift in the extent of boreal forest in North Asia and Europe due to temperatures' increase is another positive impact, but it may be slowed down by forest fires [16, 30].

Box 2.3 Forests
The World Wide Fund for Nature (WWF) states that forests represent home for 300 million people and 80% of the terrestrial biodiversity worldwide, and 1.6 billion depend on them for their livelihoods, most of them vulnerable to CC [26].

2.1.4 Water

The world already suffers from water shortage due to increased droughts [31]. Almost 3 billion people around the world suffer from water scarcity at least 1 month out of every year. Around 1.2 billion people or almost one-fifth of the world's population lack access to clean drinking water, and 500 million people are approaching this situation [32]. Figure 2.7 shows physical and economic water scarcity map around the world. The map indicates that the most suffering countries are in the MENA region and South Asia and most of Africa suffer from economic water scarcity.

Europe is the least continent to suffer from water scarcity. Only 10% of the European countries suffer from water scarcity, but the percentage will increase in the future due to an increase in water shortage [33]. Water stress will be also

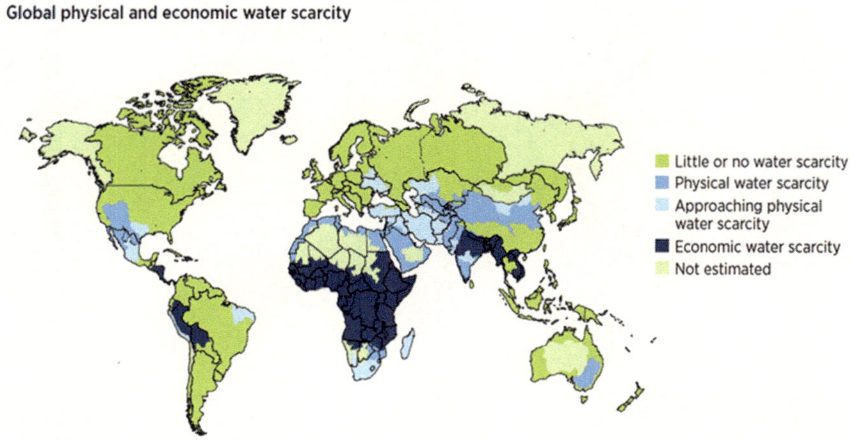

Fig. 2.7 Physical and economic water scarcity map. (Source: World Water Development Report 4, 2012)

increased in North and Latin America. Meanwhile, Africa and Asia already suffer from water stress. By 2025, it is estimated that around 5 billion people, out of a total population of around 8 billion, will suffer from water stress [34]. Water runoff changes due to change in river flows as glaciers recede and precipitation patterns change. Changes in water runoff due to glaciers recede already threat all continents except Africa [5]. Water runoff and snowmelt on mountains represent 50% of the world's freshwater [35]. Climate change will decrease freshwater falling as snow in the winter, which was stored for many months and then released to rivers in summer due to temperature rise, decreased snow falls and increased melt glaciers [36]. Simulation models expect a 10–40% increase in runoff in eastern equatorial Africa, at high latitudes and in some wet tropical areas in North America, Europe and Asia and 10–30% decrease in runoff in Southern Africa, Southern Europe, the Middle East and midlatitude western North America by the year 2050 [37]. Due to climate change, annual precipitation is projected to increase in some regions, which may seem as a positive indicator, but depends on the readiness of the infrastructure [38]. Global warming increases the severity of droughts (Fig. 2.8) as increased surface warming led to the expansion of drought areas and declination of summer precipitation in Asia and Europe. Water sources will become intermittent or disappearing in some places in Africa and North America which will hinder development in Africa. Also, water quality and quantity will be affected negatively from CC [12]. Therefore, most of the climate change impacts on water are negative 72% of total impacts, and only 28% are positive (Fig. 2.9) that demonstrate the seriousness of the negative impact of CC on water.

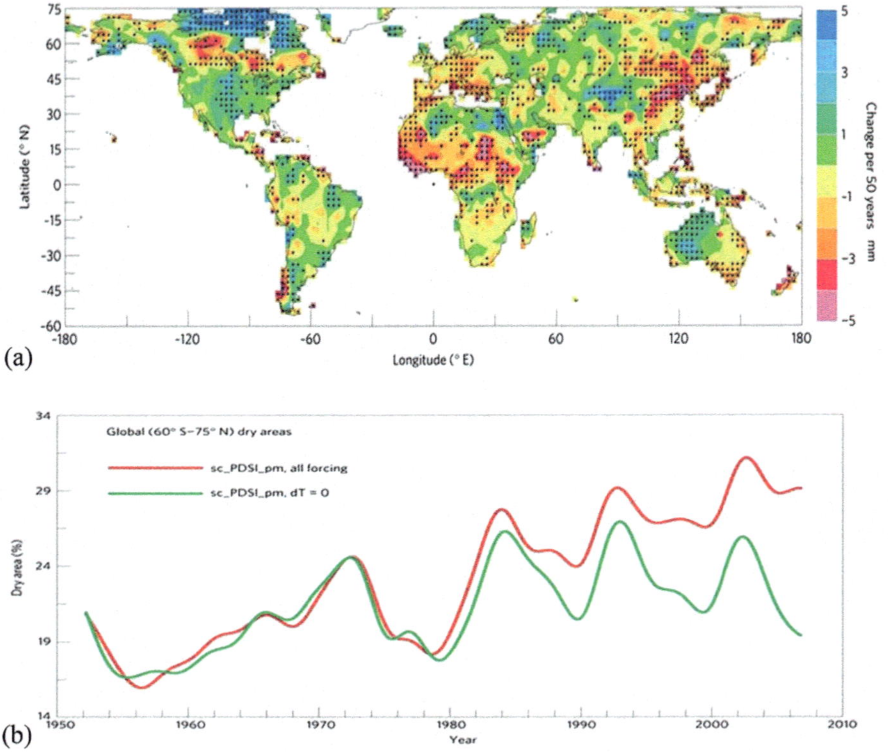

Fig. 2.8 Climate change – global annual rainfall and droughts. (**a**) Annual precipitation changes from 1950 to 2010. (**b**) Global land areas and observed surface warming (1950–2010). (Source: Aiguo Dai, Nature Climate Change 2013

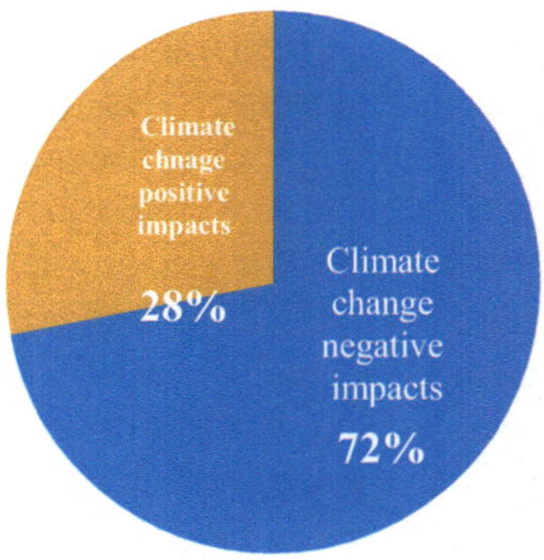

Fig. 2.9 Percentage of observed climate change negative impacts on water. (Source: Developed by authors)

Box 2.4 Water

The United Nations indicates that 2.8 billion people around the world suffer from water scarcity at least 1 month out of every year. Around 1.2 billion people or almost one-fifth of the world's population lack access to clean drinking water, and 500 million people are approaching this situation [32]. By 2025, around 5 billion people, out of a total population of around 8 billion, will suffer from water stress due to droughts [34].

2.1.5 Health

According to the World Health Organization (WHO) , weather-related natural disasters have tripled globally since the 1960s, causing 60,000 deaths every year, with most in developing countries. It is projected that between 2030 and 2050, there will be approximately 250,000 additional deaths per year due to malnutrition, malaria, diarrhoea and heat stress [39]. Also, WHO indicates 40–120 deaths per million will occur in Africa due to climate change (Fig. 2.10).

Figure 2.11 shows CC impact on human health, including higher temperature and increased frequencies of heat waves, which will increase fires and heat shocks such as the 2003 heat wave in Europe, flood and the increased risk of tropical cyclones or flood-related deaths, injuries, infectious diseases and landslides in Asia [21], North [16] and Latin America [5] and extreme temperature, floods, forest fires, cyclones and windstorms killing almost 100,000 people and affecting over 1.2 billion people from 1992 to 2001 [40].

High temperatures will change the patterns of the occurrence of disease vector risk to health in all continents; will increase transmission of disease vectors spread by mosquitoes, including malaria, dengue fever, meningitis and cholera in Africa; and will increase endemic morbidity and mortality due to diarrheal disease and cholera in Asia [41]. Global warming and floods will also increase incidence and geographic range of water-borne diseases in Africa and North America and are already a threat in Asia and Europe [42].

Food shortage causing malnutrition in developing countries due to a very difficult access to sufficient and nutritious food resulted from a decline in global crop production [43]. A decrease in annual cold-related mortality is a positive impact. According to CDC's National Centre for Health Statistics in the United States, the

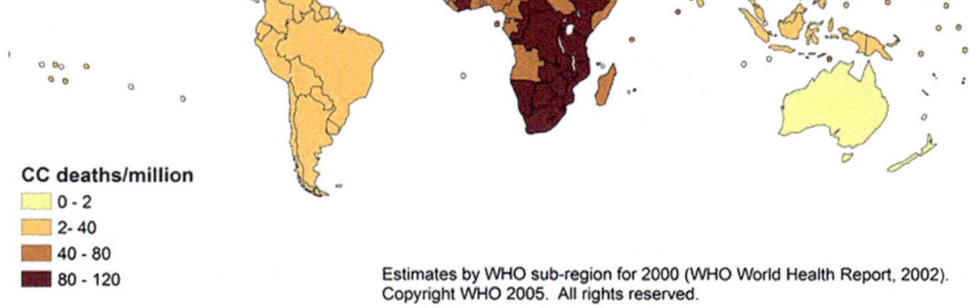

Fig. 2.10 Estimated deaths attributed to climate change. (Source: World Health Organization)

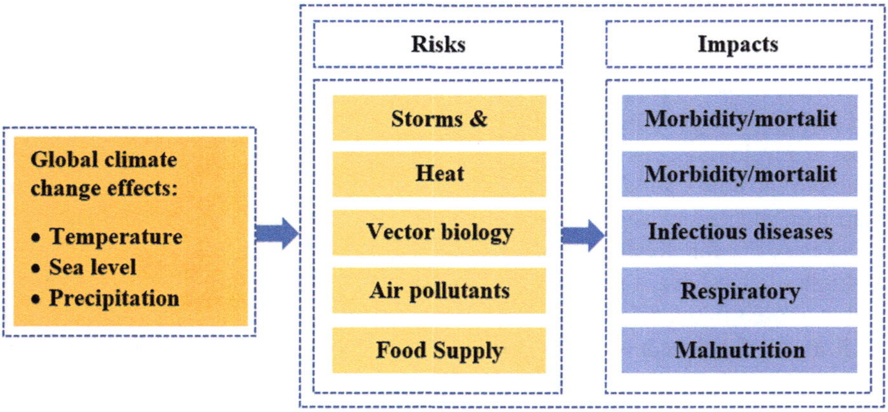

Fig. 2.11 Climate change impacts on human health. (Source: Department of Health, Tennessee, USA)

Fig. 2.12 Percentage of deaths attributed to diseases related to climate change impacts. (Source: United Nations)

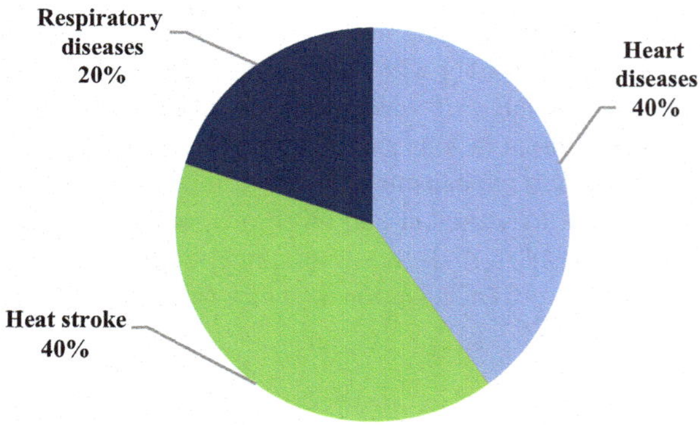

death cases caused by climate change impacts are 63%, 31%, and 6% due to heat stress, floods, and storms respectively [44]. Based on the United Nations reports, estimated deaths attributed to climate change are 40–80 deaths/million in Egypt (Fig. 2.10): 40% due to heat stroke, 40% due to heart disease and 20% due to respiratory disease as shown in Fig. 2.12.

Box 2.5 Health
According to the World Health Organization (WHO), weather-related natural disasters have tripled globally since the 1960s, causing 60,000 deaths every year.

2.1.6 *Coastal Zones*

A rise in temperature by 1 °C could have dangerous effects on the ecosystems. The temperature increase is expected to exceed 1 °C causing losses of biodiversity, ocean acidification, global warming and decreasing oxygen levels, which will act with sea level rise, causing disappearance of low-lying corals [45] in Asia, Europe and North America, and is already a threat to Africa and Latin America [16].

Coastal wetlands and mangroves create effective buffer zone to protect shoreline from erosion, storm surge and potentially small tsunami waves, which lead to property damage and deaths [46]. Coastal wetlands and mangroves will be at risk due to projected sea level rise in all continents and are already a threat for mangroves in Latin America [47]. According to the Intergovernmental Panel on Climate Change (IPCC), the observed sea level has increased at a rate of 1.7 ± 0.3 mm/year in the twentieth century [48] while, the projected sea level rise (SLR) in 2100 will be 0.5–1.4 m above the 1990 sea level (Fig. 2.13) [49]. This increase will cause inundation along coasts in Africa and North America. Also, soil salinization due to salt water intrusion will harm groundwater and agriculture areas near shores in South America [50].

Glacial melt, due to global warming, will cause floods and effect water availability and biodiversity [51]. Tropical cyclones in Asia coastal area will increase in intensity and cause damages to lives, property, crops and infrastructure due to the changes in sea surface temperature and wind flows [52]. Also, flash floods are likely to increase in Europe because warmer temperatures will increase intensity of heavy rainstorms [53].

Box 2.6 Coastal Zones
According to the Intergovernmental Panel on Climate Change (IPCC), the observed sea level has increased at a rate of 1.7 ± 0.3 mm/year in the twentieth century [48] while, the projected sea level rise (SLR) in 2100 will be 0.5–1.4 m above the 1990 sea level [49].

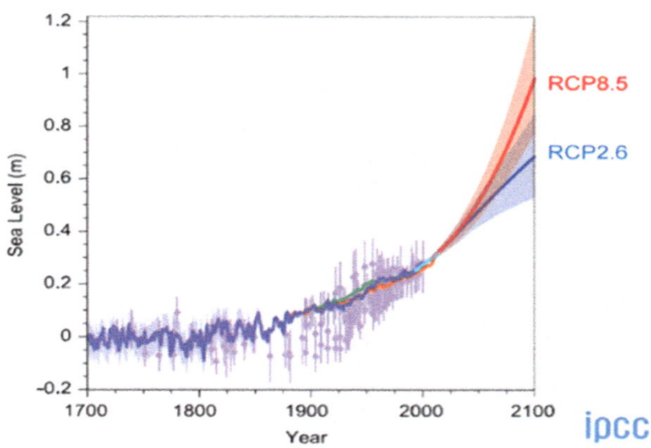

Fig. 2.13 Sea level between 1880 and 2013 and sea level projection by 2100 due to two scenarios. (Source: Jonathan Gregory, IPCC)

2.1.7 Tourism

Tourism and recreation are important economic activities. Travel and tourism contribute to 9.6% of GDP of the global economy in 2014. It is forecasted to rise by 3.7% in 2015 and projected to be 10.5% of GDP in 2025. Additionally, 103,069,000 million is the total of labour in the tourism sector and recreation sector, which is equal to 3.4% of global total employment [54]. Tourism is a source of foreign exchange in 46 out of the 49 least developed countries [55]. Climate change impacts tourism directly and indirectly. Direct impacts are on the tourism and recreation areas, while indirect impacts are on the infrastructure, which are in connection with tourism activities [56]. Figure 2.14 shows CC impact on tourism; sea level rise will threat coastal tourism, cause loss of recreational beach facilities, contribute to more acidic oceans and threat eco-tourism due to coral reef bleaching and cause loss of attractive biodiversity globally. Cities and urban tourism centres will be also at risk due to heat events, water shortages and flooding [57]. Rising temperatures will cause economic losses to winter sport industry due to snow cover reduction in Europe and Latin America, and forest fires in Africa will threat wildlife areas and parks tourism [58].

Some positive impacts of CC are that summer tourism is likely to shift to the Northwest European regions because the weather will become comfortable, unlike Southern Europe which will not be a preferred place for tourists in summer due to high temperatures. Summer tourism will shift to spring and autumn at Mediterranean countries in Europe instead of summer due to high temperatures in summer [59].

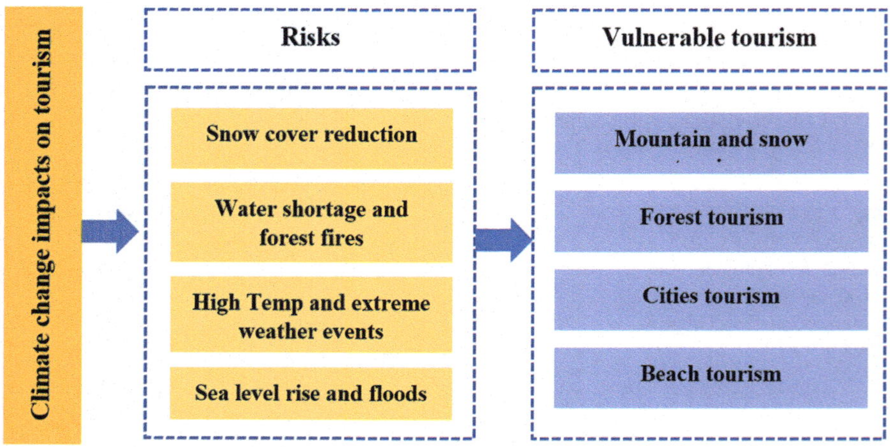

Fig. 2.14 CC risks on tourism sectors. (Source: Developed by authors)

Box 2.7 Tourism
According to the World Travel and Tourism Council (WTTC), travel and tourism contributes 9.6% of GDP to the global economy in 2014, and it is projected to be 10.5% of GDP in 2025, 103,069,000 million is the total of tourism and recreation employment, it is equal to 3.4% of global total employment [54], and tourism is a source of foreign exchange in 46 of 49 least developed countries [55].

2.1.8 *Energy*

Biomass is an important energy source for cooking and heating; dry wood and some plants represent the sources of biomass energy. Negative impact of CC on the growth and productivity of agriculture and forestry will have an effect on the availability of forest biofuels in Africa and North America and increase runoff of the river hydroelectricity potential in North America due to predicted increase of water flow rate [38, 60]. In the United States, run-of-the-river hydropower could provide about 13.7% of the total energy use in 2011, if permanently available. Another negative impact will be an increase in future energy demand for groundwater irrigation in parts of Europe and North America due to decreased rainfall. It is predicted that low availability of cool water will impact the thermal power plant efficiency [11].

According to the World Bank, water constraints are presently impacting the energy sector around the world. In Brazil, China, France, India, Sri Lanka and the United States, several power plants have lower power generation rate or are going to shut down due to low water flow or high water temperature that threatens the cooling processes during heat waves or droughts [61].

The International Energy Agency (IEA) projects that the world energy consumption will grow by 56% between 2010 and 2040 (Fig. 2.15). Most of this growth will come from non-OECD (Organisation for Economic Co-operation and Development) countries, where demand is driven by strong economic growth [62]. Aggregate commercial energy demand for cooling in Africa, Asia and Europe will increase to adapt to heat islands, which will increase heat stress in Africa, Asia and Europe [63], 1.5–2.0% increase in cooling electricity demand for every 0.6 °C increase in air temperature. From 20 °C to 25 °C, 5–10% of electricity demand used to compensate for the heat island effect [64]. Temperature rise will shift the peak electricity demand in some locations in Europe and North America from winter to summer due to a decrease in winter heating energy and an increase in summer cooling energy [19]. According to a study by the US Environmental Protection Agency, it projected an increased demand for cooling and a decreased demand for heating in most of the states in the United States by 2080–2099 (Fig. 2.16).

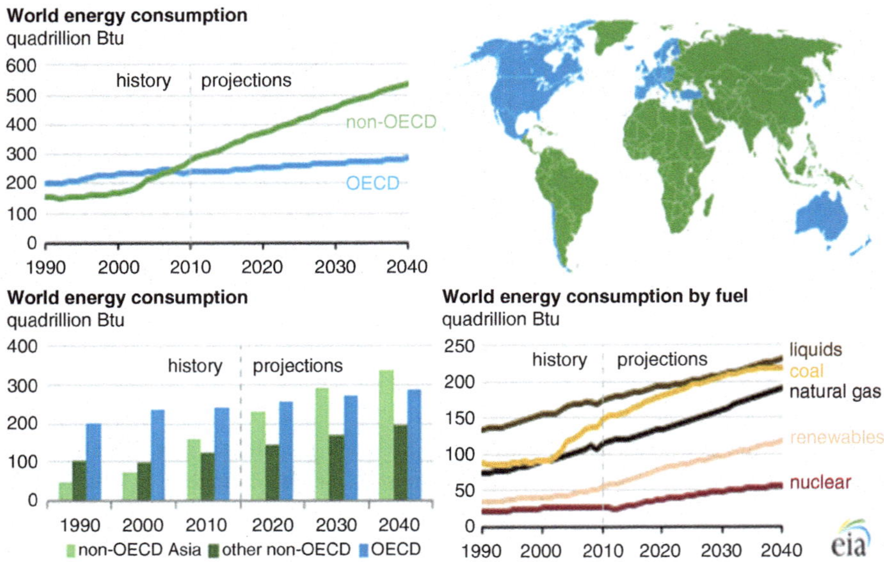

Fig. 2.15 World's energy consumption. (Source: U.S. Energy Information Administration)

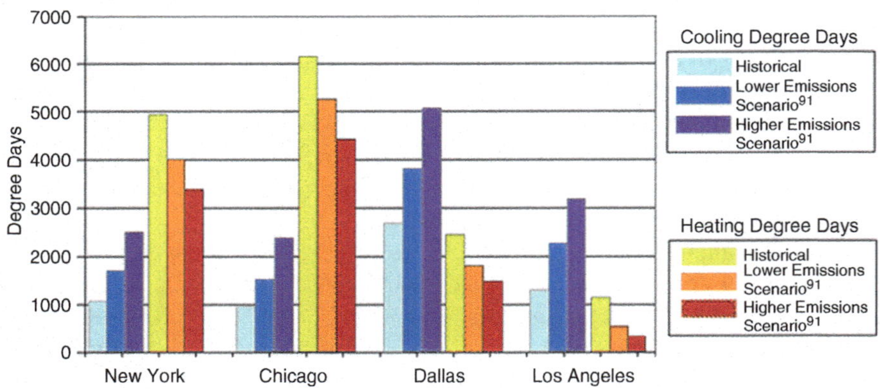

Fig. 2.16 Increase demand for cooling and decrease demand for heating in the United States by 2080–2099. (Source: US Environmental Protection Agency)

2.1.9 Fishing

Ocean acidification, high temperature and low oxygen content in seas cause changes in species range. Some of the marine ecosystems will face extinction risks. Climate change impacts will push the marine ecosystems to shift towards the north to adapt to the new changes [65]. The shift will cause an increase in fish productivity at northern waters and a decrease in productivity in the lower and midlatitudes [66]. A loss of ecosystems and fishery stocks will arise in North and Latin America and has already occurred in Africa due to CC [5]. The decline of fish stocks impacts nearly three billion people who rely on fish as a major source of protein and 12% of the world's population whom depend on fish and aquaculture to ensure their livelihoods and will affect an economic benefit of US$ 2.9 trillion per year [67].

Box 2.8 Energy
The International Energy Agency projects that the world's energy consumption will grow by 56% between 2010 and 2040 (Fig. 2.15) and most of this growth will come from non-OECD (Organisation for Economic Co-operation and Development) countries, where demand is driven by strong economic growth [62].

Box 2.9 Fishing
The decline of fishing stocks impacts nearly three billion people who rely on fish as a major source of protein and impact 12% of the world's population who depend on fish and aquaculture to ensure their livelihoods and create economic benefits of US$ 2.9 trillion per year [67].

2.1.10 Economy

Climate change caused economic losses estimated at 700 billion US dollars, close to 1% of the global GDP and 7% for developing countries in the year 2010. By 2030, global economic losses are projected to be close to 2.1% of GDP. The most affected regions are sub-Saharan Africa, Small Island, developing states and South Asia [68]. In 2011, the most affected regions were North America and Asia and then Europe. The most prevalent risks are flood and storms [69].

According to climate change scenarios, the United States will lose $1.9 trillion annually (as per the value of the current USD change rate) in 2100. Hurricane damages will cost an estimated $12 billion and more than 120 deaths per year, grown by $422 billion and 760 deaths. The US residential real estate losses will estimate to $360 billion per year due to rising sea levels in 2100. Although the loss of increased cooling energy will be amounted as $141 billion, droughts will cost the United States $950 billion to provide water to affected areas [70].

Table 2.1 shows the global climate change risks taking effect in Africa, Asia, Europe, North America and Latin America and its impact on the built environment and economic activities in urban areas such as agriculture and green areas, ecosystems, forests, water, health, coastal and flood-prone zones as well as tourism, energy,

Table 2.1 (a) Global climate change impacts on sectors (1–6)

CC challenge sectors	Climatic variables changed	CC impacts	Continents				
			Africa	Asia	Europe	North America	Latin American
Agriculture and green areas	Temperature, rainfall, solar, radiation and evaporation	Loss of land due to desertification	☑ [71] [5]	☑ [21]		☑ [72]	☑ [5] [16]
		Shorter growing seasons	✓ [71] [5] [16]				
		Growing seasons may lengthen in some areas	✓ [16]			✓ [16]	
		Changes in crop suitability, growing season and the timing of cycle of agricultural crops			☑ [12]		
		Reduction in soil fertility	✓ [14]				
		Agriculture may expand in productivity in northern areas		☑ [5]	✓ [12]		
		Soil erosion	✓ [14] [16] [5]	✓ [5]			✓ [5]
		Food insecurity and increase in the number of people at risk from hunger	✓ [5] [21]	✓ [16] [5] [21]			
		Yields from rain-fed crops could be reduced	✓ [5] [16]				
		Reduced crop productivity	✓ [21]	✓ [21] [5]	☑ [21] [12] [16]	✓ [73]	✓ [21]
		Crop productivity increase in some regions			✓ [16]		✓ [5] [16]

✓ Mark means that the impact has not occurred till now but will occur in the future
☑ Coloured tick mark in square means that the impact occurred
The dark red underline CCA risk is the positive impact
CC impact in black text is the CC risk
[ ] The red number between two red bracts is the reference

(continued)

Table 2.1 (b) Global climate change impacts on sectors (2–6)

CC challenge sectors	Climatic variables changed	CC impacts	Continents				
			Africa	Asia	Europe	North America	Latin American
Agriculture and green areas (cont'd)		Increase pest and diseases damage on agricultural	✓ [21]		✓ [16]		
Ecosystems	Temperature, rainfall, radiation, sea surface and temperature	Loss of species and extinction of many plants and animals	✓ [71] [14] [5]	☑ [5]	✓ [12]	✓ [73]	✓ [5]
		Ice cap on Mountain could disappear	✓ [16]	✓ [16]			
		Rapid melting of glaciers	☑ [21]	✓ [16]	✓ [12] [16]		✓ [16]
		The frequency of rock falls increases due to mountain destabilization			☑ [16]	✓ [16]	
		Pressure for species to shift north and to higher elevations		☑ [21]		☑ [21]	✓ [16]
		<u>Spread of warm water species</u>			☑ [21]		
		Disappearance of low-lying corals and losses of biodiversity due to SLR	☑ [16] [14] [5] [21]	✓ [21] [16] [74]	✓ [12] [16]	✓ [73]	☑ [5] [16] [21]
		Wetlands or mangroves at risk due to SLR	✓ [14] [5]	✓ [16] [74] [5]	✓ [16]	✓ [16] [73]	☑ [5]
Forests	Temperature, rainfall, solar radiation and evaporation	Forest fires	☑ [5]	☑ [60]	☑ [21] [16]	☑ [73]	☑ [16]
		<u>Northward shift in the extent of boreal forest</u>		✓ [5] [16]	✓ [16] [12]		

✓ Mark means that the impact has not occurred till now but will occur in the future

☑ Coloured tick mark in square means that the impact occurred

The dark red underline CCA risk is the positive impact

CC impact in black text is the CC risk

[ ] The red number between two red bracts is the reference

Table 2.1 (c) Global climate change impacts on sectors (3–6)

CC challenge sectors	Climatic variables changed	CC impacts	Continents				
			Africa	Asia	Europe	North America	Latin American
Water	Temperature, rainfall and evaporation	Increased drought	☑ [71] [21]	☑ [21]	✓ [16]	☑ [73]	☑ [5]
		Decline in summer precipitation		✓ [16]	✓ [12]		
		Mean annual precipitation is projected to increase in some region		✓ [16]	✓ [16]		✓ [38]
		Increased water stress	☑ [71] [5] [16] [21]	☑ [16] [5]	✓ [21] [16]	✓ [73]	☑ [16]
		Water shortage	☑ [5]	☑ [21]	✓ [21]	☑ [73]	☑ [5]
		Change runoff of water	☑ [16]	✓ [16] [5]	✓ [21]	☑ [21]	☑ [5]
		Decrease in river flows as glaciers recede		☑ [16]	☑ [12]	☑ [16]	☑ [21] [16]
Health	Temperature, rainfall and humidity	Increase incidence and geographic range of vector and waterborne diseases	✓ [21]	☑ [21] [16]	☑ [12] [16]	✓ [16]	
		Increase transmission of disease vectors, as malaria, dengue fever, meningitis, etc.	☑ [71] [38] [16] [21] [5]	✓ [16] [21] [5]	✓ [21] [12] [16]	✓ [16] [21]	✓ [5]
		Increased risk of tropical cyclones		✓ [21]		✓ [16]	✓ [5]
		Food shortage causing malnutrition	✓ [21]	✓ [21]			

✓ Mark means that the impact has not occurred till now but will occur in the future
☑ Coloured tick mark in square means that the impact occurred
The dark red underline CCA risk is the positive impact
CC impact in black text is the CC risk
[ ] The red number between two red bracts is the reference

Table 2.1 (d) Global climate change impacts on sectors (4–6)

CC challenge sectors	Climatic variables changed	CC impacts	Continents				
			Africa	Asia	Europe	North America	Latin American
Health (cont'd)		Decrease in annual cold-related mortality			✓ [12] [16]		
Coastal Zones and flood prone	Temperature, rainfall, sea level, wind and pressure	Inundation along coasts due to Sea levels rise	✓ [14] [5]	✓ [16] [74] [5]	✓ [16]	✓ [16]	✓ [21] [16] [5]
		Glacial melt-related floods		✓ [5] [16]			
		Coastal erosion and Salt water intrusion	☑ [75] [50]	☑ [76] [50]	☑ [77] [16] [50]	☑ [78] [50]	☑ [79] [50]
		increase tropical cyclones and flash floods risk		✓ [5]	✓ [16]		
Tourism	Temperature, rainfall, radiation, winds and sea level	Loss of recreational beach facilities	☑ [80]	☑ [81]			
		Wildlife areas and parks, may also attract fewer tourists	☑ [58]				
		Winter sport industry is expected to experience economic losses due to snow cover reduction	☑ [82]	☑ [81]	☑ [12] [16] [83]	☑ [84]	☑ [16]
		Summer tourism is likely to shift to the North West European regions			✓ [12]		

✓ Mark means that the impact has not occurred till now but will occur in the future
☑ Coloured tick mark in square means that the impact occurred
The dark red underline CCA risk is the positive impact
CC impact in black text is the CC risk
[ ] The red number between two red bracts is the reference

Table 2.1 (e) Global climate change impacts on sectors (5–6)

CC challenge sectors	Climatic variables changed	CC impacts	Continents				
			Africa	Asia	Europe	North America	Latin American
Tourism (cont'd)		Mediterranean countries tourism is likely to decrease in summer and increase in spring and autumn			✓ [16]		
Energy	Temperature, rainfall, radiation, winds and sea level	Increase energy demand for irrigation			✓ [38]	✓ [38]	
		Increased in aggregate commercial energy demand due to heat stress	☑ [16] [63]	☑ [16] [63]	☑ [38] [63]	☑ [63]	☑ [63]
		Increase demand for cooling and decrease for heating			✓ [12] [16]	✓ [38] [73]	
		Increase of hydropower potential of run-of-the-river hydroelectricity				✓ [38]	
		Overall decline of hydropower potential	✓ [38] [60]		✓ [12] [16]		✓ [5] [16]
		Adverse impacts on power plants efficiency due to reduced availability of cooling water	☑ [61]	☑ [61]	☑ [12] [61]	☑ [73] [61]	☑ [61]
		Reduce forest biofuels due to changes in the forest growth	✓ [60]			✓ [38]	

✓ Mark means that the impact has not occurred till now but will occur in the future
☑ Coloured tick mark in square means that the impact occurred
The dark red underline CCA risk is the positive impact
CC impact in black text is the CC risk
[ ] The red number between two red bracts is the reference

Table 2.1 (f) Global climate change impacts on sectors (6–6)

CC challenge sectors	Climatic variables changed	CC impacts	Continents				
			Africa	Asia	Europe	North America	Latin American
Fishing	Temperature, rainfall, radiation, sea surface and temperature	Loss of ecosystems and fishery stocks	☑ [16] [14] [5] [21]			✓ [73]	✓ [5] [16]
		production of marine fisheries in the North Atlantic are likely to increase			✓ [16]		
Economy	Temperature, rainfall, sea level, wind, pressure, solar radiation and evaporation	Increased economic losses	☑ [61]	☑ [61]	☑ [61] [12]	☑ [61]	☑ [61]

✓ Mark means that the impact has not occurred till now but will occur in the future
☑ Coloured tick mark in square means that the impact occurred
The dark red underline CCA risk is the positive impact
CC impact in black text is the CC risk
[ ] The red number between two red bracts is the reference

fishing and economy. Also, the table shows the changed climatic variables such as temperature, rainfall, sea level, wind, pressure, solar radiation and evaporation. This table is based on a literature review of many studies and reports regarding the investigation of the geographical distribution of climate change impacts. The table's importance lies in the arrangement of continents in terms of exposure to and classification of the CC impacts into positive and negative impacts; results of this table are summarized in Table 2.2.

In Table 2.1, the dark red underlined texts refer to the positive impacts, and normal black texts refer to the negative impacts (CC risk). The tick mark (✓) means the impacts have not occurred till now, but CC scenarios project will occur in the future, while blue-coloured tick mark in a square (✓) means that the impacts occurred. The red number between two bracts ([n]) is the reference.

Table 2.2 classified global climate change risks (CC) among positive or negative CC impacts in order to investigate the advantages and disadvantages of CC and answer the following two questions:

- Is the overall climate change impact harmful or beneficial?
- Which continent is the most vulnerable to climate change?

Table 2.2 Classification of global climate change risks

Classification of global CC Risks	Africa	Asia	Europe	North America	Latin American	Total
Positive impact already occurred	0	1	1	0	0	2
Future positive impacts	1	2	7	2	2	14
Total of positive CC impacts	**1**	**3**	**8**	**2**	**2**	**16**
Negative impact already occurred	17	15	11	12	14	69
Future negative impact	15	15	17	14	10	71
Total of negative CC impacts	**32**	**30**	**28**	**26**	**24**	**140**
Total of CC impacts	**33**	**33**	**36**	**28**	**26**	**156**

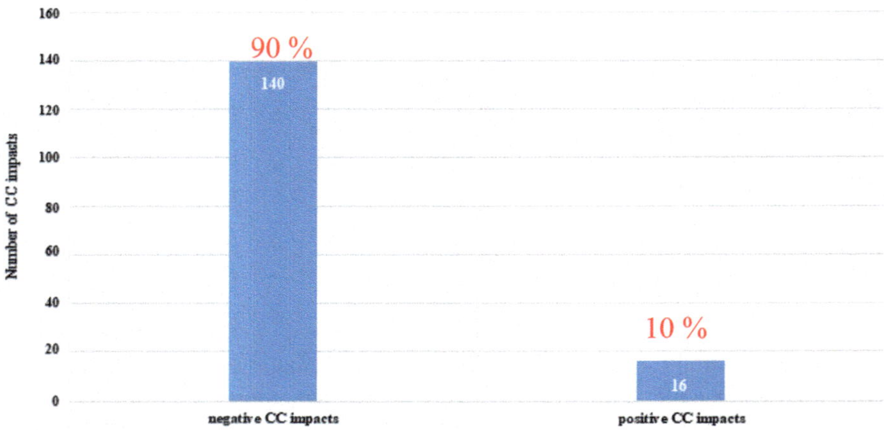

Fig. 2.17 Classification of climate change impacts. (Source: Developed by authors)

Climate change positive impacts represent only 10% of the total impacts where the 90% are negative impacts (Fig. 2.17). Only 2 out of 16 positive impacts of CC have already occurred, whereas approximately half of the negative impacts of CC have already occurred (Fig. 2.18). Figure 2.19 shows the continents' share of climate change risks, by descending order Africa, Asia, Europe, North America and Latin America. However, the descending order of the positive impacts of CC on continents is Europe, Asia, North and Latin America, and at last Africa, which indicates that most of CC positive impacts will occur in Europe, but most of CC negative impacts occurred in Africa (a huge number of risks), which is estimated at 141

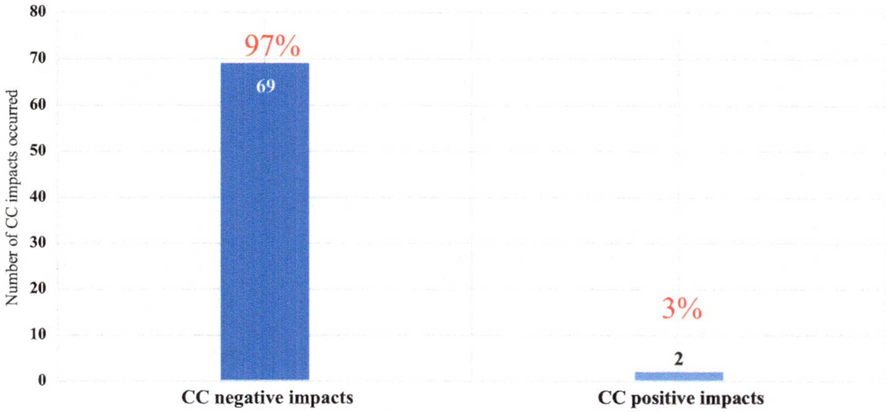

Fig. 2.18 Classification of climate change impacts occurred. (Source: Developed by authors)

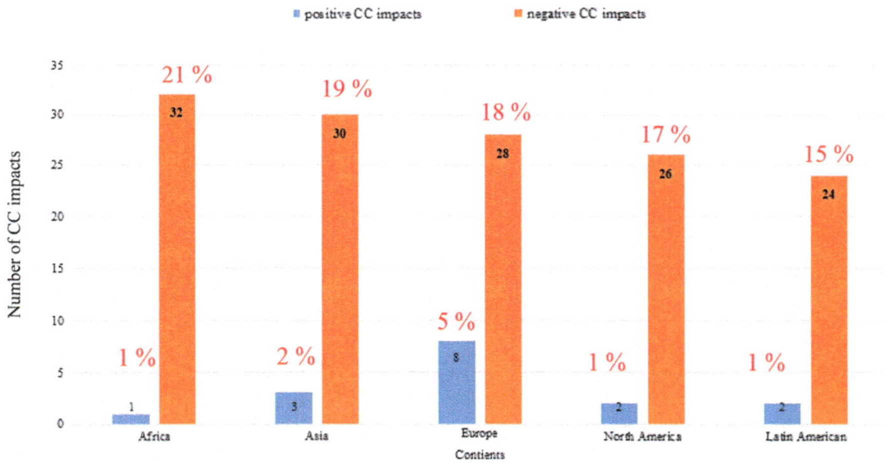

Fig. 2.19 Share of continent from climate change impacts. Europe, largest number of positive impacts & Africa largest number of negative impacts. (Source: Developed by authors)

risks' threats. There are risks impacting different sectors in comparison to modest positive impacts as shown in Fig. 2.20.

Based on the above, it is concluded that most of the climate change impacts are negative and that Africa is the most continent to suffer from CC.

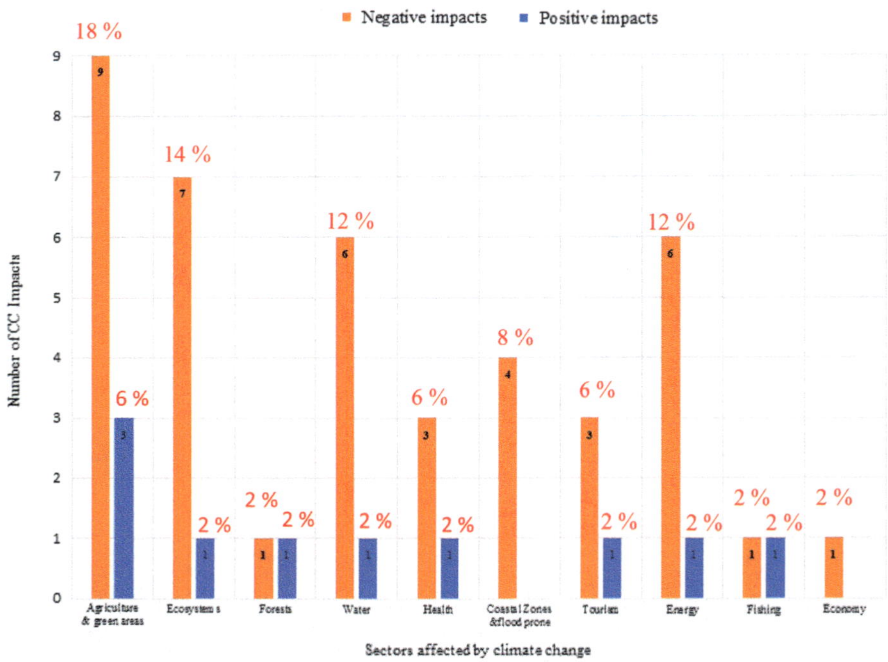

Fig. 2.20 Classification of climate change impacts by sectors. (Source: Developed by authors)

Box 2.10 Climate Change Adaptation Challenges

1. Agriculture

   - The most expected impact to occur on agriculture is crop productivity reduction, but the most occurred risk is the loss of land due to desertification. All of these impacts mean food insecurity, which will eventually lead to an increase in food prices.
   - The positive impact is the increase of crop productivity in some regions, and agriculture will expand in productivity in northern areas of Asia and Europe.

2. Water

   - Most of the world will suffer from water shortage due to increased water stress and changing in water runoff. Also, a low water quality due to pollutions and floods poses risks on drinking water.
   - Water shortage means loss in most of sectors such as agriculture, health and settlements, industry and energy, because water is one of the basic needs for life.

3. Forest
   - Wildfires cause losses in the ecosystem, property loss, human mortality and air pollution.
4. Coastal areas
   - Sea level rise (SLR) will cause submergence of wetlands and mangroves, coastal flooding and coastal erosion. All of these risks will lead to the disappearance of corals, losses of biodiversity and waves of migration as well as urban damage and economy loss in coastal and low-lying areas.
5. Health
   - Heat stress, floods, fires, food, water-borne diseases and vector-borne diseases will lead to deterioration in health status, injuries and deaths, especially for the elderly and the poor.
   - Under-nutrition will increase due to food shortage.
   - Positive effects include a decrease cold-related mortality and morbidity in some areas due to fewer extreme cold.
6. Ecosystem
   - Rapid melting glaciers and ice cap on mountains disappear as a result of global warming, which will damage boreal-tundra Arctic.
   - Boreal-tundra Arctic, Amazon forest and marine ecosystem organisms will face imbalance in the ecosystem, which will constitute a major threat to organisms.
7. City and infrastructure
   - Vulnerable and the labour class, who live in poor quality housing and infrastructure, will strongly suffer from climate change impacts.
8. Energy
   - Increase in the load on the electricity grids as a result of the increased energy demand in urban areas due to global warming.
9. Policy
   - Conflicts will occur over water resources, food and energy globally.

2.2 Conclusion

This chapter focuses on assessing the potential global impacts of climate change (CC) through analysing the literature review of investigating CC consequences on the sectors associated with the built environment. The review was based on identifying vulnerability, weighing the negative and positive consequences and the status of impacts in terms of occurrence. This task has assisted in establishing a substantial and thorough understanding of the potential global impacts of CC and investigating their footprint on the built environment. It has been noticed that there are a colossal number of CC risks, which is estimated to be 141 impacts distributed globally; 90% of them are negative, and only 10% are positive impacts. It is important to mention that 13% of CC positive impacts have occurred; however, approximately half of the CC negative impacts have already occurred. Africa and Asia are the most exposed continents to CC risks. However, it is important to notice that there are slight differences between the continents in terms of the type of risks. The most positive impacts present in the high-latitude regions of Europe, but it is modest in comparison with other risks.

Results indicate the serious risks of climate change, the few positive impacts and the increasing rate of risks over time affecting all countries, especially developing ones. Climate change negatively affects all sectors. For example, in agriculture sector, it will cause soil erosion and pest spreading resulting in food insecurity. In forests, it will cause high temperature and droughts accelerating forest fires. As far as ecosystems sector is concerned, it will cause wild fires or melting ice leading to species extinction or migration to adapt. For the water sector, a decrease in water runoff will affect water quantity and quality. Heat stress will increase transmission of disease vectors. SLR, coral beaching and snow cover reduction will impact ecotourism. In the energy sector, global warming will increase the energy demand, water shortage will affect hydropower potential and reduced cooling water will lower thermal power plant efficiency. Thus these risks will turn into global economic losses and reflect on the built environment.

References

1. EPA, "Frequently Asked Questions: Back to Basics" United States Environmental Protection Agency, April 2009.
2. Quaye W, Yawson RM, Ayeh ES, "Climate Change and Food Security: The Role of Biotechnology" *African Journal of Food*, vol. 12, 2012.
3. Intergovernmental Panel on Climate Change (IPCC), "Climate Change 2014: Impacts, Adaptation and Vulnerability, Part B: Regional Aspects" Cambridge University Press, 2014.
4. Ulrich Hoffmann, "Agriculture at the crossroads: assuring food security in developing countries," in *Trade and Environment Review 2013*, United Nations Conference on Trade and Development, 2013, pp. 2–8.
5. UNFCCC, "Climate Change: Impacts, Vulnerabilities and Adaptation in Developing Countries," United Nations Framework Convention on Climate Change (UNFCCC), 2007.

6. Jim Ritter, "Soil Erosion – Causes and Effects," OMAFRA Factsheet, 2012 10. [Online]. Available at: www.omafra.gov.on.ca/english/engineer/facts/12-053.htm. (Accessed: June 1, 2015).
7. Alberto Marini and Mohamed Talbi, Desertification and Risk Analysis Using High and Medium Resolution Satellite Data, Springer Science & Business Media, 2008, pp. 274.
8. Ebenezer Olagunju, "Drought, desertification and the Nigerian environment" *Journal of Ecology and the Natural Environment*, vol. 7, pp. 196–209, 2015.
9. UNESCO, "What is desertification? Objective: Introduce the phenomenon of desertification" [Online]. Available at: www.unesco.org/mab/doc/ekocd/chapter1.html.(Accessed: June 2, 2015).
10. E.C. Oerke, "Crop losses to pests," The Journal of Agricultural Science, vol. 144, no. 01, pp. 31–43, 2006.
11. Hugh Turral, Jacob Burke and Jean-Marc Faurès, Climate change, water and food security, Rome: Food and Agriculture Organization of the United Nations (FAO), 2011.
12. Sara Venturini, Silvia Medri and Sergio Castellari, "Overview of key climate change impacts, vulnerabilities and adaptation action in Europe" Climate Services Division (SERC), 2012.
13. Carlo Lavalle et al, "Climate change in Europe, Impact on agriculture and forestry" Agronomy for Sustainable Development, Springer, 2009.
14. UNFCCC, "Background paper on Impacts, vulnerability and adaptation to climate change in Africa" in *African Workshop on Adaptation Implementation of Decision 1/CP.10 of the UNFCCC Convention*, Accra, Ghana, 21–23 September, 2006.
15. R.J. Thomas, "Opportunities to reduce the vulnerability of dry land farmers in Central and West Asia and North Africa to climate change" *Elsevier*, vol. 126, no. 1–2, pp. 36–45, 2008.
16. The Intergovernmental Panel on Climate Change (IPCC), "Climate Change 2007: Impacts, Adaptation and Vulnerability" Cambridge University Press, 2007.
17. Economic and Social Development Department, "Climate change and agriculture: physical and human dimensions," [Online]. Available at: www.fao.org/docrep/005/y4252e/y4252e15.html. (Accessed: June 3, 2015).
18. Jørgen E. Olesena and Marco B. Tjele, "Consequences of climate change for European agricultural productivity, land use and policy" *European Journal of Agronomy*, 20 January 2002.
19. Roman Römisch, "Regional Challenges in the Perspective of 2020" Ismeri Europa, 2009.
20. Mark C. Urban, "Accelerating extinction risk from climate change" *the American Association for the Advancement of Science (AAAS)*, vol. 348, no. 6234, pp. 571–573, 2015.
21. The Intergovernmental Panel on Climate Change (IPCC), "Climate Change 2014: Impacts, Adaptation and Vulnerability" Cambridge University Press, Cambridge, UK, 2014.
22. National Academy of Sciences- Institute of Medicine, "Ecological Impacts of Climate Change", National Research Council, 2008.
23. European Environment Agency, "Vulnerability and adaptation to climate change in Europe" European Communities, Copenhagen, 2006.
24. K. M. Brander, "Global fish production and climate change" *The National Academy of Sciences of the USA*, September 26, 2007.
25. San Francisco State University, "Climate change could cause cold-blooded animals' thermal tolerance to shrink" *Science Daily*, May 19, 2015.
26. "OVERVIEW" WWF's, [Online]. Available at: http://www.worldwildlife.org/habitats/forest-habitat. (Accessed: March 19, 2016).
27. Andrei P. Kirilenko and Roger A. Sedjo, "Climate change impacts on forestry" *National Academy of Sciences*, vol. 104, 2007.
28. Craig D. Allena, Alison K. Macaladyb et al, "A global overview of drought and heat-induced tree mortality reveals emerging climate change risks for forests", Elsevier, October 21, 2009.
29. World Watch Magazine, "A Burning Issue" *World Watch Magazine*, vol. 20, no. 4, August 2007.
30. Roger Olsson, "Boreal Forest and Climate Change - Regional Perspectives," the Air Pollution & Climate Secretariat, May 2010.

31. "Climate Impacts on Water Resources," US Environmental Protection Agency, [Online]. Available at: http://www3.epa.gov/climatechange/impacts/water.html. (Accessed: May 5, 2016).
32. "Water Scarcity | International Decade for Action 'Water for Life' 2005–2015", October 20, 2013. [Online]. Available at: www.un.org/waterforlifedecade/scarcity.shtml. (Accessed: June 6, 2015).
33. "Water Scarcity & Droughts in the European Union", European Commission, April 22, 2015, [Online]. Available at: ec.europa.eu/environment/water/quantity/about.htm. (Accessed: June 8, 2015).
34. Nigel W. Arnell, "Climate change and global water resources" *Elsevier*, October 1999.
35. Water Program, GRACE Communications Foundation, "The Impact of Climate Change on Water Resources" GRACE Communications Foundation [Online]. Available at: www.grace-links.org/2380/the-impact-of-climate-change-on-water-resources. (Accessed: June 10, 2015).
36. Hamlet A. Fluharty et al, "Effects of Climate Change on Water Resources in the Pacific North West: Impacts and Policy Implications" in *Climate and Water Policy Meeting*, University of Washington, 2001.
37. P. C. D. Milly, K. A. Dunne & A. V. Vecchia, "Global pattern of trends in stream flow and water availability in a changing climate" *Nature*, vol. 438, pp. 347–350, November 17, 2005.
38. IPCC, "Climate Change 2001: Impacts, Adaptation and Vulnerability" The Inter-governmental Panel on Climate Change (IPCC), 2001.
39. World Health Organization, "Climate change and health" [Online]. Available at: http://www.who.int/mediacentre/factsheets/fs266/en/. (Accessed: March 19, 2015).
40. United Nations Environment Programme, "GEO4 environment for development Summary for Decision Makers" UNEP, 2007.
41. Owen C. Owens et al, "Climate change and Health Across", United Nations Economic Commission for Africa, 2011.
42. "HUMAN HEALTH" Climate Institute, [Online]. Available at: http://www.climate.org/topics/health.html. (Accessed: May 3, 2016).
43. Oyiga Benedict et al, "Implication of Climate Change on Crop Yield and Food Accessibility in Sub-Saharan Africa" University of Bonn, Bonn, 2011.
44. Jason Samenow, "Cold kills more than heat, CDC says; other researchers caution not necessarily" Washington Post, August 4, 2014.
45. Victor S. Kennedy et al, "Coastal and Marine Ecosystems & Global Climate Change: Potential Effects on U.S. Resources" Pew Center on Global Climate Change, 2002.
46. Keryn B. Gedan et al, "The present and future role of coastal wetland vegetation in protecting shorelines" *Publisher Springer Netherlands*, vol. 106, no. 1, pp. 7–29, 2011.
47. Mario D.P. Godoy and Luiz D. de Lacerda, "Mangroves Response to Climate Change: A Review of Recent Findings on Mangrove Extension and Distribution" *Anais da Academia Brasileira de Ciências*, vol. 87, no. 2, 2015.
48. John A. Church and Neil J. White, "A 20th century acceleration in global sea-level rise" *Geophysical Research Letters*, vol. 33, no. 1, January 6, 2006.
49. Stefan Rahmstorf, "A Semi-Empirical Approach to Projecting Future Sea-Level Rise", Science, Vol. 315, pp. 368–370, December 14, 2006.
50. Swagata Guha, "Variable-Density Flow Models of Saltwater Intrusion in Coastal Landforms in Response to Climate Change Induced Sea Level Rise" University of South Florida, Graduate School, 2010.
51. Jianchu Xu, et al, "The Melting Himalayas: Cascading Effects of Climate Change on Water, Biodiversity, and Livelihoods" *Society for Conservation Biology*, pp. 520–530, MAY 15, 2009.
52. Doo-Sun R Park, Chang-Hoi Ho and Joo-Hong, "Growing threat of intense tropical cyclones to East Asia over the period 1977–2010" *IOP Publishing*, January 15, 2014.
53. European Commission Climate Action, "How will we be affected, floods, droughts, landslides and other effects", European Commission Climate Action, May 8, 2015. [Online]. Available at: ec.europa.eu/clima/policies/adaptation/how/index_en.htm. (Accessed: June 12, 2015).

54. The World Travel & Tourism Council, "WTTC Travel & Tourism Economic Impact 2015" Oxford Economics Ltd, [Online]. Available at: www.hospitalitynet.org/file/152005348.pdf. (Accessed: March 12, 2016).
55. World Tourism Organization Network, "FAQ - Climate Change and Tourism" [Online]. Available at: http://sdt.unwto.org/content/faq-climate-change-and-tourism. (Accessed: 26 June 2015).
56. Alain Safa and Nathalie Hilmi, "The impact of climate change on coastal tourism in MENA countries" *Topics in Middle Eastern and African Economies*, vol. 14, September 2012.
57. Mark Nicholls, "Climate Change: Implications for Tourism, Key Findings from the Intergovernmental Panel on Climate Change Fifth Assessment Report" University of Cambridge, May 2014.
58. Isaiah Esipisu, "Kenya tackles climate threats to wildlife" Thomson Reuters Foundation, Jan 2013.
59. Bas Amelung, Krzysztof Blazejczyk and Andreas Matzarakis, "Climate Change and Tourism" 2007.
60. Hany Besada and Nelson K. Sewankambo, "Climate Change in Africa: Adaptation, Mitigation and Governance Challenges" CIGI Special Report.
61. Diego J. Rodriguez, "Interlinkages and trade-offs between water and energy," The World Bank, 2014.
62. "International Energy Outlook 2013" International Energy Agency - IEA, 2013.
63. Shushi Peng et al, "Surface Urban Heat Island Across 419 Global Big Cities" Environmental Science & Technology, 2011.
64. Akbari, Hashem, "Energy Saving Potentials and Air Quality Benefits of Urban Heat Island Mitigation" Lawrence Berkeley National Laboratory, 2005.
65. Scott C. Doney et al, "Climate Change Impacts on Marine Ecosystems annual Review of Marine Science" *Web of Science Journals*, vol. 4, pp. 11–37, January 2012.
66. M. Barange et al, "Impacts of climate change on marine ecosystem production in societies dependent on fisheries" *Nature*, 2014.
67. FAO, "The State of World Fisheries and Aquaculture" Food and Agriculture Organization of the United Nations, Rome, 2014.
68. DARA and the Climate Vulnerable Forum, "Climate Vulnerability Monitor 2nd Edition, guide to the Cold Calculus of a Hot Planet" Fundaci6n DARA Internacional, 2012.
69. The World Economic Forum, "Global Risks Report 2013", The World Economic Forum, 2013.
70. Frank Ackerman and Elizabeth A. Stanton, "The Cost of Climate Change, What We'll Pay if Global Warming Continues Unchecked" Natural Resources Defense Council, May 2008.
71. "Impacts of Climate Change," TERI (The Energy and Resources Institute), [Online]. Available at: http://climateofconcern.org/impacts-of-climate-change/. (Accessed: March 3, 2016).
72. Michelle Schwartz et al, "Desertification and Migration: Mexico and the United States" Natural Heritage Institute, San Francisco, California, 1994.
73. United States Environmental Protection Agency, [Online]. Available at: https://www.epa.gov/climate-impacts/climate-change-impacts-sector. (Accessed: January 10, 2017).
74. Mahfuz Ahmed and Suphachol Suphachalasai, "Assessing the Costs of Climate Change and Adaptation in South Asia", Asian Development Bank, June 2014.
75. Isabelle Niang, "Coastal erosion major threat to West Africa," February 13, 2012. [Online]. Available at: http://www.rtcc.org/2012/01/26/coastal-erosion-major-threat-to-west-africa/. (Accessed: July 15, 2015).
76. "The Economics of Climate Change in Southeast Asia", Asian Development Bank, 2009.
77. "East Riding coastal erosion" June 24, 2014. [Online]. Available at: www.theguardian.com/environment/gallery/2014/jun/24/east-riding-coastal-erosion-in-pictures. (Accessed: July 14, 2015).
78. "US West Coast Erosion Spiked In Winter 2009, Previewing Likely Future As Climate Changes" U.S. Department of the Interior, U.S. Geological Survey, December 7, 2011. [Online]. Available at: www.usgs.gov/newsroom/article.asp?ID=2849from=rss_home. (Accessed: July 18, 2016).

79. Nelson Guillermo et al, "Coastal erosion along the Caribbean coast of Colombia: Magnitudes, causes and management", *Elsevier*, vol. 114, 2015.
80. A.C. Ibe and L.F. Awosika, "Sea level rise impact on African coastal zones" African Centre for Technology Studies, Lagos Nigeria, 1991.
81. Y.-P. Su and C. Michael Hall, "Chapter 2 Climate Change and Tourism in Asia", UNWTO, 2013.
82. Smart N. Uchegbu and Ejikeme J. Kanu, "The Impact of Climates Change on Tourism in Africa" Department of Urban and Regional Planning, University of Nigeria, Enugu State, Nigeria.
83. Elsasser, H and Burki, R, "Climate change as a threat to tourism in the Alps" *Climate Research*, no. 20, pp. 253–257, 2002.
84. Elizabeth Burakowski and Matthew Magnusson, "Climate Impacts on the Winter Tourism Economy in the United States" Natural Resources Defense Council, 2012.

Chapter 3
Climate Change Impacts on Urban Areas and Infrastructure

Abstract In recent decades, climate change (CC) impacts on human settlements have been manifested worldwide. Such impacts arise from CC risks, notably sea level rise (SLR), desertification, drought, extreme events, food security, increased health risks and temperature-related morbidity in urban environments. Existing trends towards urbanization have been witnessed due to the negative impacts of climate change on rural livelihoods. However, settlement patterns' changes may not only be restricted to the socially vulnerable but also will be extended to coastal areas' dwellers and SLR, which will force people to migrate to unaffected areas. Large population movements caused by deteriorated environmental conditions may lead to conflict through competition for resources in the receiving area. It may also contribute to the failure of infrastructure and services' provision. This chapter investigates the direct and indirect impacts of climate change on urban areas and infrastructure. The indirect impacts are reflected on the built environment and consequently on the economy of urban and rural regions. Moreover, the chapter explores whether all climate change risks occurred or not. Finally, it determines the continents that are most exposed to climate change risks related to the built environment and, thus, addresses the direct and indirect impacts on the built environment with a wider view about climate change consequences.

3.1 Indirect Impact of CC on Urban Areas and Infrastructure

Based on the impacts of climate change on different sectors explained in Chap. 2, this chapter investigates the consequences of climate change impacts on agriculture, ecosystems, forests, health, coastal zones, tourism, energy and economy on the built environment, which is known as *indirect impact of climate change*.

M. M. Aboulnaga et al., *Urban Climate Change Adaptation in Developing Countries*, https://doi.org/10.1007/978-3-030-05405-2_3

3.1.1 *Impact of Agriculture and Food Security Risks*

Migration of farmers to new settlements due to droughts, desertification and extreme events will increase the stress on cities infrastructure and lead to changing land use. Most of this migration will go to informal settlements, which will also increase the pressure on it [1]. In Nigeria, for example, the desertification effect can be explained as 3500 square kilometres of land is turning into desert every year, which forces farmers to move and exacerbate the problem of the unbalance between agricultural resource and population density which is a problem in the MENA region and Central Asia. As shown in Fig. 3.1, the red areas represent the affected areas [2].

3.1.2 *Impact of Ecosystem Risks*

Global warming, droughts and extreme events put pressure on species to migrate to find suitable condition. The migration of ecosystem is isolated by roads, settlements, canals and electric power lines, which block ecological migration and lead to injury to humans, property damage and loss of habitat [3]. In 2002, about 100 black bears, 1291 snakes, 1333 frogs, 374 turtles, 265 birds, 72 mammals, 29 alligators and 1 lizard were killed in NC, USA, when wildlife was crossing Florida's highway [4]. The Federal Highway Administration (FHA) reported that animal-vehicle collisions doubled between the years 1990 and 2004, from about 175,000 to

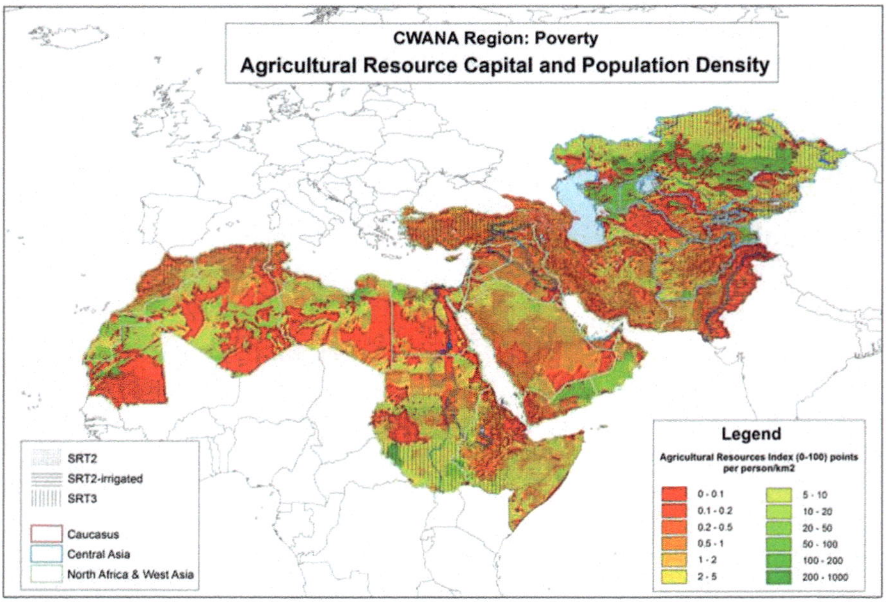

Fig. 3.1 Agricultural resource poverty to population density. (Source: Geoinformatics Solutions)

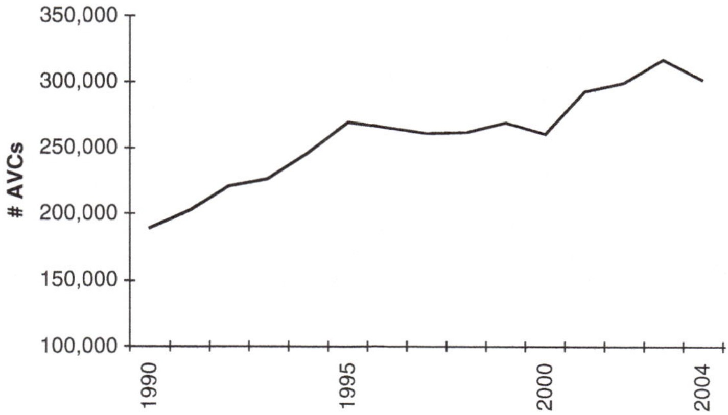

Fig. 3.2 Animal-vehicle collisions (AVCs) from 1990 to 2004. (Source: Federal Highway Administration)

Fig. 3.3 Elephant crossing the road causing threats to human's safety. (Source: Sweetsangram, Wikimedia Commons)

around 300,000 (Fig. 3.2). Humans' safety has been threatened when elephants are crossing the road as shown in Fig. 3.3. Between 1977 and 2006, wildebeest populations in Nairobi National Park, Kenya, also declined by 93.4% due to the blockage of migratory corridors by urban areas as presented in Fig. 3.4 [5]. The above indicates the possibility of aggravating the problem in the future.

Rising temperatures and melting snow cover will affect the stabilization of mountains. The frequency of rock falls and landslides will be increased due to the

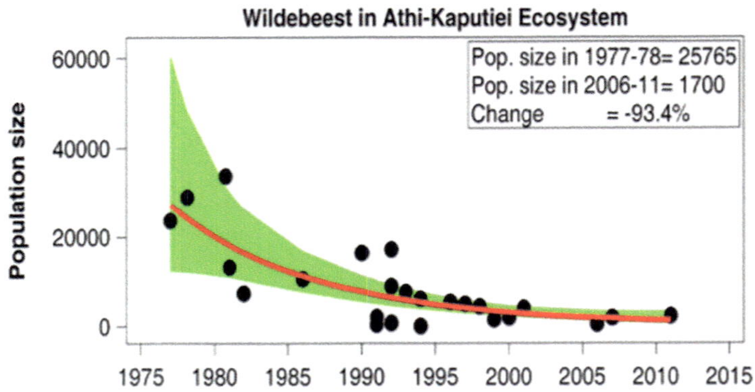

Fig. 3.4 Decline in wildebeest populations in Nairobi National Park. Africa. (Source: UNEP global environmental alert service)

Fig. 3.5 Landslides destroy buildings near mountains. (Source: Antandrus, Wikicommons)

destabilization of mountains, which will destroy settlements and infrastructure near these mountains [6]. About 24% of the northern hemisphere land surface contains permafrost. Thus, North America will be exposed to an increase in rock fall activity in the future due to climate change [7]. Every year, 25–50 people are killed by landslides in the United States alone [8]. Also, it was observed that increase in rock fall activity during the summer of 2003 in Alpine permafrost in Europe is due to increase temperature in that year [9]. Figure 3.5 shows landslides destroy buildings near these mountains.

3.1.3 Impact of Forest Risks

In the last year, forest fires that spread to settlements near forests are due to increased temperature and drier weather, namely, in California (USA), Spain and Greece. Increased spread and intensity of forest fires led to loss of lives, property and infrastructure [10]. All continents are suffering from the impacts of forest fires on settlements [11]. Forest fires increase local air pollution in cities, causing lung diseases and breathing difficulties even in healthy individuals. In the United States, financial loss due to property damages, between 2000 and 2009, is estimated to be US$665 million per year as well as loss of lives [12]. Figure 3.6 shows forest fire in a Chilean city that was responsible for killing 11 lives and the destruction of 1000 buildings. In July 2018, forest fire severely erupted in the coastal city of Mati, Greece, killing more than 90 people and injured more than 150 as well as destroyed hundreds of homes [13].

3.1.4 Impact of Water Risks

Low accessibility, quality deterioration, and demand of water in cities due to climate change impacts; such as droughts, floods and higher temperature led to water scarcity [14]. Floods decrease water quality and increase temperature which increases water demand. Slum areas or informal settlements suffer the most from low water access. In cities, the impacts of CC on water availability led to loss of lives and millions of US dollars daily (2.5 billion), and 768 million people have no access to safe sanitation and water. Experts estimate that by 2080, 43–50% of the global population will be living in water-scarce countries, compared to 28% today [15]. Most of these populations are in MENA region, horn of Africa and Central Asia. According to the World Resources Institute, water scarcity will expand to

(a) (b)

Fig. 3.6 Forest fire in the United States and South Europe. (**a**) Forest fire in a Chilean city in 2011, USA. (**b**) Forest fire in Mati city, Rafina, in 2018, Greece. (Image Source: **a**. Mrsramsey, Wikimedia Commons. **b**. AFP)

include some regions in Europe and North and Latin America [16]. Currently only 50% of Nairobi's inhabitants have no access to clean water, and 60% do not receive water constantly [17].

3.1.5 Impact of Health Risks

Heat waves affect health and decrease productivity of construction workers; besides, heat stroke leads to increased mortality and morbidity rates. Labour productivity will decrease globally by up to 20% in hot months by 2050 due to global warming [18]. A study in Chennai, India, examined the relationship between climate conditions and productivity. This study looked at the change of daily work outputs of construction workers compared to the temperature change values. It also expressed productivity loss due to temperature change, and productivity loss varied from 44 to 54 and was projected to be reduced by 80% by 2050 [19]. Another study projected a decreased productivity of construction male workers by approximately 20% and 40% in cold days and hot days, respectively (Fig. 3.7).

3.1.6 Impact of Coastal Zone and Flood Risks

Floods cause damage to infrastructure and buildings, including possible structural failure of buildings, riverine and coastal zones being the most vulnerable. The damage intensity of flood depends on flood water velocity not the flood water depth [20]. Damage to infrastructure also causes disruptions to the supplies of clean water,

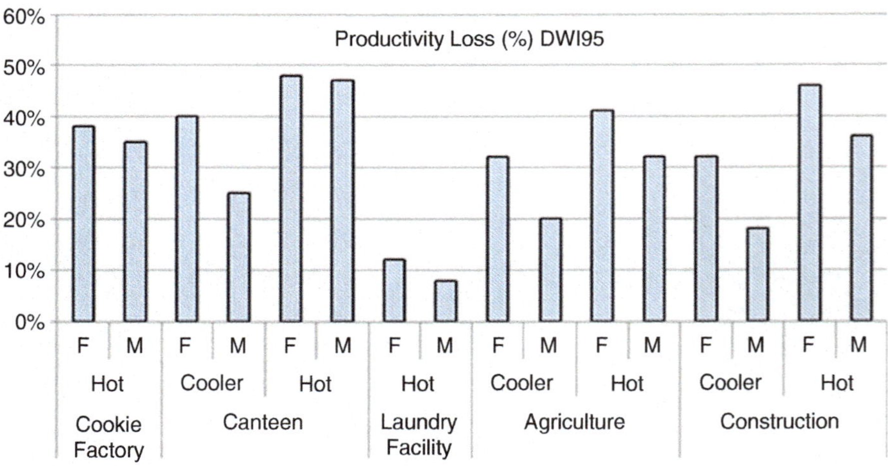

Fig. 3.7 Productivity loss due to temperature change, (F = female, M = male). (Source: Karin Lundgren, Global Health Network)

wastewater treatment, electricity, transport, communication, education and health care, which eventually leads to loss of livelihoods, injuries and deaths and land degradation. Australia's economic and financial loss due to floods over the period 1967–2005 is averagely US$377 million per year [21]. Loss from sandy storm is estimated at US$15.9 billion for New York and New Jersey due to damage of buildings and infrastructure as shown in Fig. 3.8 [22]. Coastal erosion, salt water intrusion and soil salinization will also damage the foundations of buildings and roads and affect underground water. Salinity will affect surface and subsurface reservoirs like lakes, soil moisture and groundwater due to salt water intrusion [23].

A study on coastal Bangladesh roads predicted that maintenance expenditure of paved roads will be increased by 252% due to the increase in ground water salinity. It is important to note that soil salinization causes blistering and cracking of road surfaces [24].

Tropical cyclones will damage infrastructure and buildings, especially high-rise buildings. Tropical cyclones affect zones from about 25 km to 500 km, globally. Approximately 80 tropical cyclones occur each year, causing losses estimated at billions of USD, killing about 10,000 people and causing inundation of low-lying coastal areas, erosion of coastline, saltwater and damaging buildings and transport networks in Asia, Europe and America [25].

Fig. 3.8 Settlement damage due to sandy severe storm and floods. (Source: U.S. Air Force photo by Master Sgt. Mark C. Olsen)

3.1.7 Impact of Tourism Risks

Climate change will influence tourists' flows and destinations. Mountain regions and coastal destinations are the most affected areas resulting from climate change risks. The most affected tourism segments are beaches, nature and winter sport tourism [26]. Regions depending on tourism are under the threat of sea level rise (SLR) that will submerge small islands and coastal regions.

Desertification and the scarcity of water make regions less hospitable for tourists. Deforestation also harms biodiversity, and snow melting affects ski resorts and biodiversity on the mountains. Construction and maintenance of recreational buildings will decrease due to tourism economic loss [27].

3.1.8 Impact of Energy Risks

Heat island effect increases energy consumption for cooling in summer. In 1800, only 3% of the world's population lived in cities, whereas in 2005, 50% lived in cities and consume over 75% of the world's energy use. By 2030, it is predicted that 60% will reside in urban areas [28]. Heat island effect costs the city of Los Angeles about 100 million US$ per year in the energy sector [29].

A study in 30 urban areas in Athens, Greece, to evaluate the impact of the urban climate on the energy consumption of buildings, found out that for the city of Athens, where the mean heat island intensity exceeds 10 °C, the cooling load of buildings may be doubled and the peak electricity load for cooling purposes may be tripled [30]. The Environmental Protection Agency projected an increase cooling degree day and decrease heating degree days from 2005 to 2050 due to global warming as shown in Fig. 3.9 [31].

The study analyses the impacts of global change on the hydropower potential of Europe in the future, project reductions in hydropower potentials in southern and south-eastern European countries due to low river flow as a result of glaciers recede and low precipitation pattern [32]. The study also analyses the future impact of global change on Batoka Gorge hydroelectric power station in Zimbabwe. Reductions in electricity production on Batoka Gorge hydroelectric power station are projected as a result of a reduction in the flow of the Zambezi river [33].

There is a decrease in thermal power plant efficiency due to reduced availability of cooling water, during droughts and heat waves. The nuclear power plant production loss may exceed 2% per degree Celsius. In 2004, a study found out that increase in ambient temperature as it happens in the desert environment, decrease thermal efficiency by 3–8% and reduce base load plant capacity and output by 20–24% for a gas-powered plant. Foreseen overall decline of hydropower potential and decreased thermal power plant efficiency due to reduced availability of cooling water will increase energy outages in cities and raise cost of buildings materials due to energy

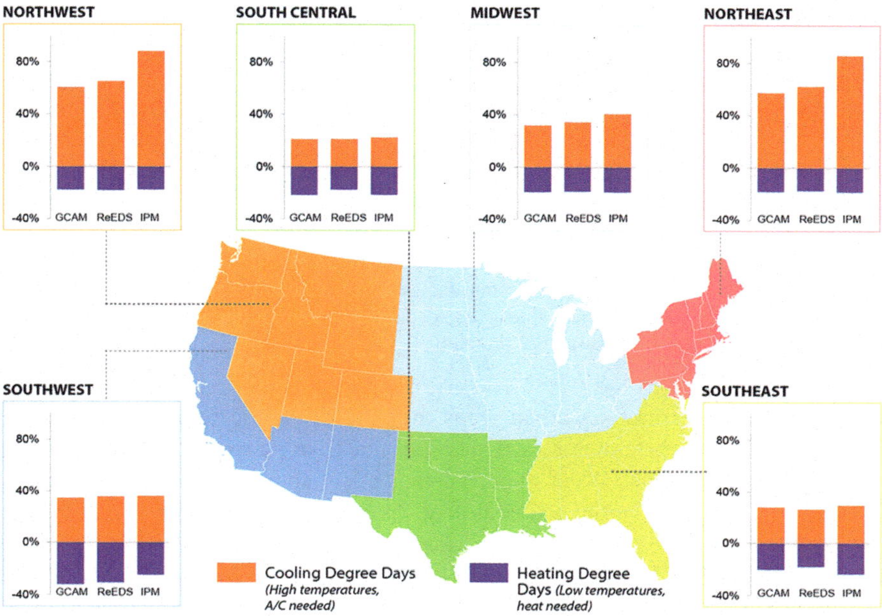

Fig. 3.9 Projected impact of CC on US heating and cooling degree days (2005–2050). (Source: Environmental Protection Agency)

shortage [34]. A decreased winter heating energy will be followed by an increased summer cooling energy, so the overall energy consumption will be increased [35].

3.1.9 Impact of Economy Risks

People spend high proportion of their incomes on basic needs such as housing, energy and food, which are expected to experience the hardest impacts. People will suffer from poor housing quality, which will increase health impacts, resulting in higher morbidity and mortality rates [36]. Low income people living in informal settlements are the most vulnerable to CC impacts, especially in developing countries. They often live in most exposed areas to the effects of global climate change with little or no infrastructure existing to provide protection from extreme events or to ensure mobility. Informal settlements have no buildings' regulations and lack housing finance. Low-quality housing will have weak resistance to floods, global warming and extreme events [37]. About 1.6 billion people out of 6.5 billion, which represents about 25% of the total world population, live in substandard housing, and 100 million are homeless in 2005. This represents about 25% of the world's total urban population, and the number of slum dwellers worldwide will increase by 2030 to nearly 2 billion [38].

Table 3.1 Climate change impact on real estate sector (after Sven Bienert [41])

Climate aspect	Commercial and residential real estate
Rise in temperature	Reduced ground rent (lower potential revenue due to regional population changes; also, increased need for cooling and thus higher operating costs)
Water scarcity	Decline in attractiveness of a region/decline in ground rent; higher costs for water supply and treatment
Rising sea level	Reduced settlement area in coastal regions
Increase in extreme weather events	Direct loss (e.g. hail damage to buildings) Indirect loss (e.g. through gaps in production or rent after hurricanes) Consequential loss (e.g. declining number of tourists in flood areas and rising insurance premiums)

Climate change risks on urban areas increase economic losses of real estate investment. The rise in temperature reduced ground floor rent due to higher operating costs of cooling systems. Water scarcity led to a decline in ground floor rent due to higher costs for water supply and treatment. Extreme weather events and rising sea level affect real estate prices and rise insurance premiums [39]. There is a relation between adaptive buildings to climate change and real estate value, 4.8% higher rents for energy efficiency (Energy Star) buildings compared to non-Energy Star in USA [40]. According to Urban Land Institute, CC impacts affect property value by rise in temperature and water scarcity as it led to reduced buildings rent due to higher operating costs. Rising sea level reduces settlement area in coastal regions. An increase in extreme weather events causes direct loss in building damages, indirect loss through gaps in production or rent after hurricanes and consequential loss through declining number of tourists in flood areas and rising insurance premiums as presented in Table 3.1 [41].

3.2 Direct Impact of CC on Urban Areas and Infrastructure

Around half of the world's population live in urban areas and are expected to increase by 60% by 2030. Urban areas are affected by heat island due to global warming. An urban heat island effect (UHIE) is described as the warmth of both the atmosphere and surfaces in cities compared to rural surroundings [28].

A rise in energy production to meet the increased energy demand for cooling due to UHIE will increase air pollution and GHG emissions from power plants, transport and waste [42]. Urban heat island affects health, especially on the vulnerable such as children, the elderly and patients with respiratory diseases. It also causes general discomfort and heat stroke. In 1979–2003, more than 8000 deaths were due to heat stroke in the United States. This number exceeded the normal recorded figures of mortalities by extreme weather events [43]. Also, high temperatures of pavements' surfaces, specifically asphalt and dark colour tiles, led to an increase in the surfaces' temperatures; besides, the temperature of sewage water temperature,

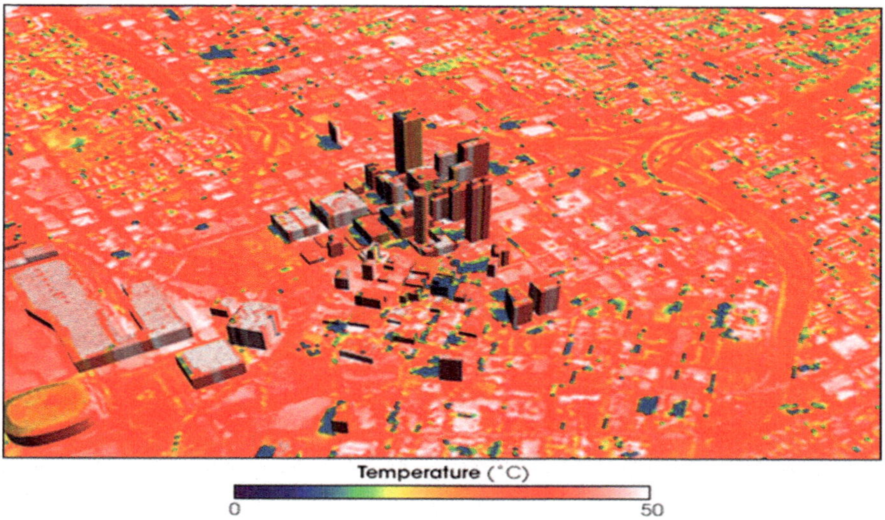

Fig. 3.10 Image of Atlanta, Georgia, showing temperature distribution. (Source: Ryanjo, en. wikipedia)

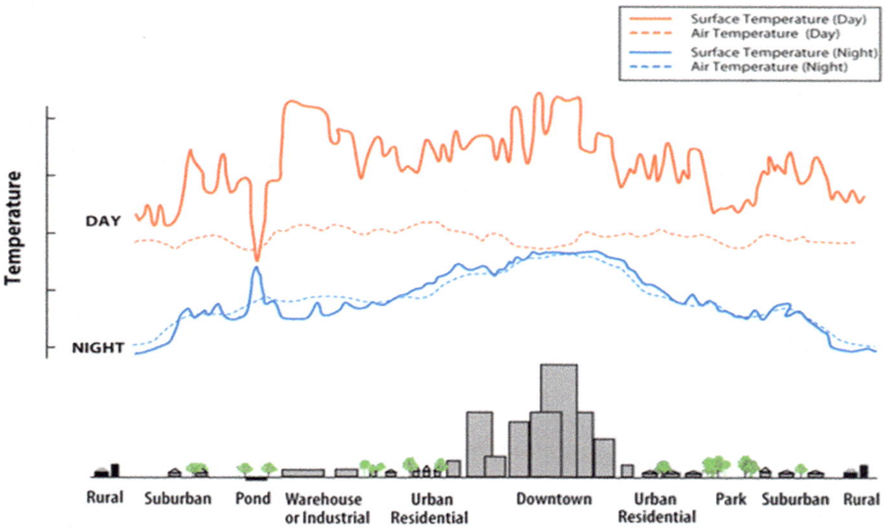

Fig. 3.11 Surface and atmospheric temperatures of different land use area. (Source: U.S. Environmental Protection Agency)

underneath these surfaces, exceeded 35 °C. Consequently, when such hot water is released into rivers and lakes, this will raise the water temperature and harm aquatic life [44]. Figures 3.10 and 3.11 show that the urban temperatures are lower at the urban-rural border and parks than in dense urban areas.

Climate change will force many people to immigrate in the next 40 years. Large displacement will occur within developing countries. The impact of CC on migration will become clearly tangible when global temperature rises 2 °C. Migration may also increase violent conflicts, while adaptation policy will minimize the risks. Therefore, climate change risks will force communities to migrate away from the risks' zones. In addition, droughts force some pastoralists (sheep or cattle farmers) from the Sahel (north of Egypt) and Sudan to permanently migrate to safer areas. River banks' erosion in Bangladesh, land degradation in southern Tanzania and droughts in in northern Ethiopia forced people to migrate. Climate change will force 200 million people to migrate by 2050 [45]. The UN Refugee Agency (UNRA) estimated that approximately 24 million people have migrated due to environmental factors between 2002 and 2012. Also, 141 million people lost their homes due to 3559 natural hazard events from 1980 to 2000 and 97% of the affected people lived in developing countries. Sea level rise (SLR) is threatening 41% of the world's population living within 100 Km of the coast. The SLR may displace more than 14 million Egyptians in 2050 [46]. According to Internal Displacement Monitoring Centre (IDMC), the percentage of displaced persons due to drought in northern Kenya, southern Ethiopia and south-central Somalia is projected to reach 30% by 2040 (Fig. 3.12). Floods and storms played a major global role in regional displacement between the years 2008 and 2013 (Fig. 3.13). In 2013, 80% of the 20 largest events took place in Asia, with typhoons, floods and earthquakes. Two of the largest displacements of 2013 occurred in the Philippines in September by a typhoon that displaced 1.7 million people and in November displaced 4.1 million [47].

An increase in weather-related disasters around the world will lead to an increase in forced migration in the future. Destination cities will suffer from an increased pressure on infrastructure, housing, medical and social services and population increase. All this pressure will lead to more waste management problems resulting

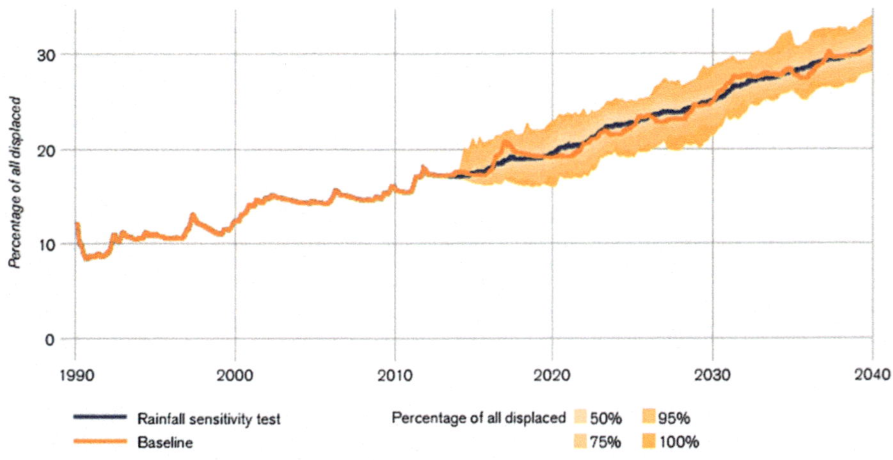

Fig. 3.12 Percentage of displaced due to drought in northern Kenya, southern Ethiopia, and south-central Somalia. (Source: Internal Displacement Monitoring Centre (IDMC))

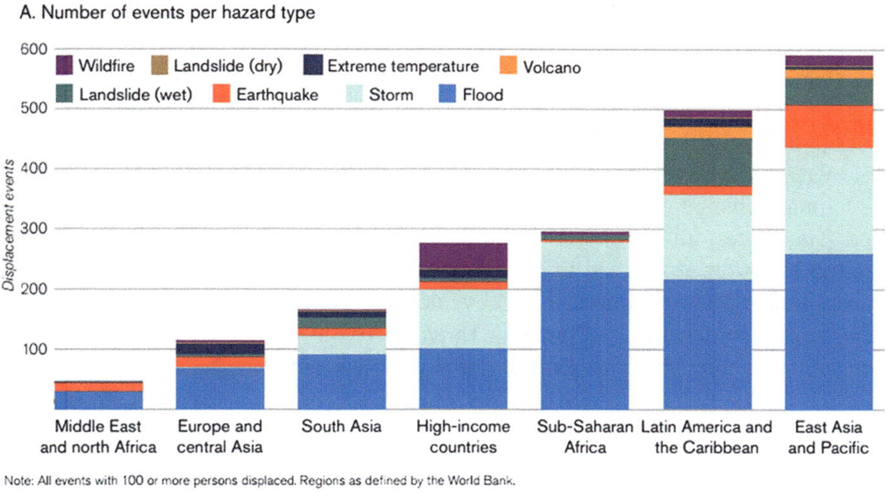

Fig. 3.13 Regional displacement by hazard type. (Source: Internal Displacement Monitoring Centre (IDMC))

in widespread public health threats. In addition, displaced persons who cannot find adequate accommodation will be forced to build their own makeshift shelter in slums and shanty towns. Over 80% of internally displaced families in Khartoum are living in temporary shelters made out of plastic and paper. Almost 90% of shelters are vulnerable to extreme natural events. In Japan, the Kobe earthquake displaced 300,000 people, and more events such as the eruption of Mount Pinatubo volcano in the Philippines, hurricane Katrina and the tsunami in Sri Lanka increase the risk.

Most of the recent environmental disaster caused massive internal displacements [48]. Around 1.3 million Somalis were internally displaced due to drought in 2011, and 290,000 people are seeking refuge across international borders due to Horn of Africa drought [49]. Slums and congestion will increase electricity outages and force stress on sewage systems, especially in developing countries due to already overstretched infrastructure [50].

Recently, climate change has been causing internal displacement due to an increased drought, desertification, salinization of groundwater and soil and rising sea levels. In the future, CC will force some groups to migrate across international frontiers to neighbouring countries to avoid the risk. Yet, these groups will face problems while migrating because none of the existing international refugee law instruments interact with environmental disaster refugees [51]. Climate change also increases power outages by flood, high winds and other extreme events. California storm caused power outages effect on 113,000 households, closed roads and caused the cancellation of more than 200 flights [52]. High winds led to power cuts in 300 homes in Dorset, UK, and trees falling causing numerous roads to be blocked [53]. Las Conchas wildfire in New Mexico caused a threat to the power grid that delivers electricity to about 400,000 customers in summer 2011. Power outages by extreme events as Hurricane Sandy led to large financial loss in the United States, which cost the nation between US$27 billion and US$52 billion dollars in 2012 [54]. Also, weather caused 80% of all outages between the years 2003 and 2012, and 59% of weather-related outages were caused by storms and severe weather (Fig. 3.14) [55]. In September 2017, the recent storm in Florida left 6.5 million homes without electricity after Hurricane Irma that cut a deadly path through the state [56].

Climate change causes disruption of infrastructure due to flooding, landslides and fallen trees; all are caused by high wind, storm surge and floods. In Torbay, United Kingdom floods led to closure of the Torbay road four times in February causing massive disruption to 17,000 vehicles. Falling trees and landslides caused 250 roads' incidents [57]. Figures 3.15 and 3.16 show New Orleans, USA, after floods that submerged approximately 80% of the city [58].

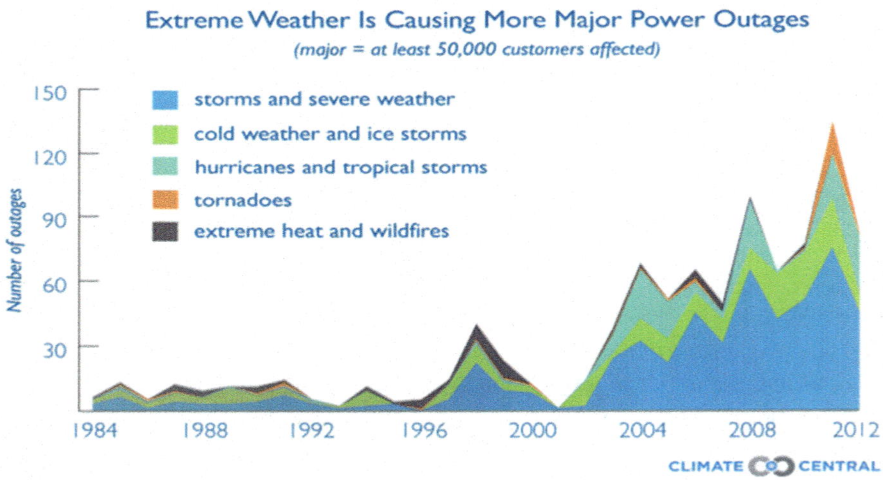

Fig. 3.14 Weather-related blackouts in the United States. (Source: Climate Central)

Fig. 3.15 Severe storm 'Sandy' and 'Katrina' in New Orleans, Louisiana, USA. (Source: https://commons.wikimedia.org/wiki/File:FEMA_-_15012. Photo credit: Jocelyn Augustino)

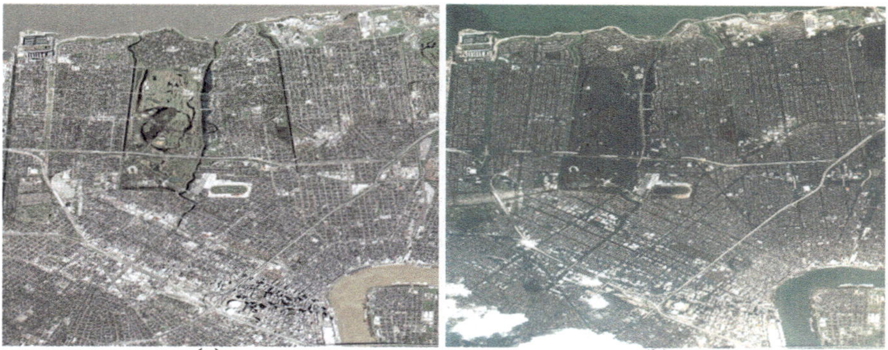

(a) (b)

Fig. 3.16 Satellite photos of New Orleans, USA, after (right) and before (left) floods. (**a**) Before floods (**b**) After floods. (Source: WikiCommons)

Table 3.2 shows weather-related disasters around the world in 5 years between 2010 and 2015. The table indicates the intensity and severity of weather-related disasters that resulted in huge losses in life and economy.

Area of forests and timber prices will increase in some areas and decrease in others. Scenarios project that the overall impacts of CC on timber markets will be beneficial [71]. On the other side, there are other scenarios that project an increase in the global timber prices. These price changes are due to an increase in forest fires that will hinder expanding forests northward.

Forests will shift northern due to increase in precipitation and temperature causing reduced winter snow pack, which will increase global forest area by 5–6% by 2050. Forest productivity is also expected to increase, and timber harvests will increase 6% in 2050. This increase may lower average timber prices [72, 73].

Table 3.3 shows global CC risks on agriculture and green areas, ecosystems, forests, water, health, coastal and flood-prone zones, tourism, energy and economy that indirectly affect cities and buildings in Africa, Asia, Europe, North America and Latin America. The table is based on literature review of many studies and reports for investigating the geographical distribution of global CC risks and their effect on cities and buildings.

This table is important in order to classify the continents in terms of the exposure of the built environment to CC.

Results of Table 3.3 are summarized in Table 3.4. The CC risks in row header refer to the risks and their effect on cities and buildings identified and concluded in row header titled CC risks impacts on construction sector.

In Table 3.3, the dark red underlined text refers to a positive impact, while the black text refers to a negative impact (climate change risk).

The tick mark (✓) means that the impact has not occurred yet, but the CC scenarios project that it will occur in the future. Also, the coloured tick mark (☑) in the blue square means that the impact has occurred, whereas the red number between two bracts ([n]) is the reference.

Table 3.2 Weather-related disasters around the world (1–2)

Risk	Country	Date	Loses
Floods	United States	2017	134 deaths [59]
Heat wave	India	2016	160 deaths [60]
Floods	France	2015	20 dead [61]
Floods	South-eastern Africa	2015	200 people dead and 120,000 forced from their homes [61]
landslide	Burma	2015	100 dead [62]
Floods	Alexandria, Egypt	2015	Seven dead [63]
Heat wave	India- Pakistan	2015	2500 dead in India and 1229 people in Pakistan [61]
Landslide	India	2014	209 dead [64]
Floods	Nepal	2014	241 dead [64]
Landslide	Afghanistan	2014	256 dead [64]
Cold wave	Peru	2014	505 dead [64] (2–2)
Floods	India, Pakistan	2014	665 dead [64]
Sandy storm	U.S., Caribbean, Bahamas	2013	254, 65.00 billion loses [65]
Heat wave	India	2013	531 dead [66]
Floods	India	2013	1537 dead, 4211 missing 271,931 homeless 1.1billion loses [66]
Floods	Nepal	2013	118 dead, 1 missing 6 injured 4314 homeless [66]
Floods	Pakistan	2013	234 dead 93,000 homeless [66]
Storm surge	Philippines, Vietnam, China, Palau	2013	7345 dead [66]
Flooding	North Korea, South Korea	2012	1901, 1.04 billion loses [65]
Flooding	Nigeria	2012	363, 636 million loses [65]
Typhoon	Philippines, Palau	2012	1901, 1.04 billion loses [65]
Drought/heat wave	United States	2012	20.00 billion loses [65]
drought	East Africa	2011	30,000 dead [67]
Floods	Thailand	2011	657 dead and $45 billion loss (18% of the country's GDP) [67]
Tornado	US Southeast	2011	321 dead, $7.3 billion losses [67]
Floods	Columbia	2011	116 died and $5.85 billion loses (2% of Columbia GDP) [67] (2-2)
Storm	Philippines	2011	1249 died [67]
Storm	France	2010	51 dead [68]
Typhoon	Philippines	2010	31 dead [69]
Floods	Mexico	2010	1000 dead [70]
Wildfires	Russia	2010	15,000 deaths and $15 billion loss [70]
Floods	Pakistan	2010	2000 dead and over 20 million affected, 9.5$ billion losses [70]

Table 3.3 (a) Global climate change risks and their impact on cities and buildings (1–4)

CCA challenge sectors	CC risks impacts on construction sector	CC risks	Continents				
			Africa	Asia	Europe	North America	Latin America
Agriculture and green areas	Migrate framers into new settlements to seek new livelihoods that increase stress on cities [74]	Loss of land due to due to droughts, desertification and extreme events	☑ [75]	☑ [75]		☑ [76]	☑ [77]
Ecosystems	Migration of ecosystem isolated by roads and settlements will harm settlements [78] ' [79]	Pressure for species to shift north and to higher elevations Ice cap on Mountain could disappear	☑ [59]	✓ [80]	✓ [81]	☑ [78]	✓ [80]
	Landslides will destroy roads and buildings near mountains [6]	The frequency of rock falls will increase due to destabilization of mountains			☑ [9]	✓ [7]	
Forests	spread fires to settlements near forests [12]	Forest fires	☑ [11]	✓ [82]	☑ [11]	☑ [12]	☑ [83]
Water	decrease Water access in cities And increase in Water quality problems and water demand	Water shortage	☑ [84]	☑ [17]	✓ [85]	✓ [85]	☑ [85]

✓ Mark means that the impact has not occurred till now but will occur in the future

☑ Coloured tick mark in square means that the impact occurred

The dark red underline CCA risk is the positive impact

CC impact in black text is the CC risk

[ ] The red number between two red bracts is the reference

Table 3.3 (b) Global climate change risks and their impact on cities and buildings (2–4)

CCA challenge sectors	CC risks impacts on construction sector	CC risks	Continents				
			Africa	Asia	Europe	North America	Latin America
Health	Decrease productivity of construction workers [86]	Risk of Heat waves on health.	✓ [87]	✓ [18]	✓ [23]	✓ [87]	✓ [87]
Coastal Zones and Flood Prone	Disruption of buildings and infrastructure due to floods [25]	Inundation along coasts due to Sea levels rise or Flash floods	☑ [88]	☑ [89]	☑ [90]	☑ [91]	✓ [92]
	Will damage foundations of buildings and roads [24]	Coastal erosion and Salt water intrusion soil salinization	☑ [93]	☑ [94]	☑ [95]	☑ [96]	☑ [97]
	Contamination of water supply [23]						
Tourism	Economic loss to recreational buildings sector	Loss of recreational beach facilities Effects on winter tourism due to snow cover reduction	☑ [85]	☑ [98]	☑ [85]	☑ [99]	☑ [100]
Energy	Consume a lot energy for cooling	heat island effect	☑ [101]	☑ [101]	☑ [102]	☑ [102]	☑ [102]

✓ Mark means that the impact has not occurred till now but will occur in the future
☑ Coloured tick mark in square means that the impact occurred
The dark red underline CCA risk is the positive impact
CC impact in black text is the CC risk
[ ] The red number between two red bracts is the reference

Table 3.3 (c) Global climate change risks and their impact on cities and buildings (3–4)

CCA challenge sectors	CC risks impacts on construction sector	CC risks	Continents				
			Africa	Asia	Europe	North America	Latin America
Energy (cont.)	Increase energy outages in Cities and raising of buildings materials cost due to energy shortage	Foreseen overall decline of hydropower potential Decrease thermal power plant efficiency due to reduced availability of cooling water	☑ [33] [103]	☑ [103]	☑ [32] [103]	☑ [103]	☑ [103]
Economy	Recession in real estate investment	Increased economic losses	☑ [40]	☑ [40]	☑ [40]	☑ [104]	☑ [105]
Urban Areas	Heat island effect and Declining air quality in cities	Increase temperatures	☑ [101]	☑ [101]	☑ [101]	☑ [101]	☑ [101]
	Reduction in housing quality	Temperature rise, floods and extreme weather events	☑ [101]	☑ [101]	✓ [101]	✓ [101]	☑ [101]
	Greater risks to public safety, loss of property and large displacement of people away from risk zone	Flood, rural fire, landslide and storm events	☑ [45]	☑ [45]	☑ [85]	☑ [45]	☑ [85]

✓ Mark means that the impact has not occurred till now but will occur in the future
☑ Coloured tick mark in square means that the impact occurred
The dark red underline CCA risk is the positive impact
CC impact in black text is the CC risk
[ ] The red number between two red bracts is the reference

Table 3.3 (d) Global climate change risks and their impact on cities and buildings (4–4)

CCA challenge sectors	CC risks impacts on construction sector	CC risks	Continents				
			Africa	Asia	Europe	North America	Latin America
Urban Areas (cont.)	- Power outages disruption by flood and high winds -Disruption of infra-structure due to flooding, landslides, fallen trees and lines.	Increases Intense tropical cyclone activity, flood	☑ [101]	☑ [106]	☑ [57]	☑ [54]	☑ [101]
	Increase pressure on urban areas and infrastructures in areas where displaced people will go to it [48]	Migrate into new settlements, seek new livelihoods	☑ [107]	☑ [107]	✓ [25]	☑ [107]	☑ [107]
	Increase timber production and drop timber prices	Increase forest productivity of northern sphere	✓ [73]	✓ [73]	✓ [73]	✓ [73]	✓ [73]

✓ Mark means that the impact has not occurred till now but will occur in the future
☑ Coloured tick mark in square means that the impact occurred
The dark red underline CCA risk is the positive impact
CC impact in black text is the CC risk
[ ] The red number between two red bracts is the reference

Table 3.4 shows the classification of global climate change risks on cities and buildings. The table indicates that the global CC impacts on cities and buildings are mostly negative. Figure 3.17 shows that 75% of CC risks on cities and buildings have already occurred. Risks frequency and severity gradually increase overtime; hence the next chapter will focus on CC adaptation measures to counterbalance effect of CC risks.

Till now, Africa is the most affected continent, but in the future CC risks will extend to affect all continents equally. This is mainly demonstrating that CC risk on cities and buildings affects all mankind unlike the rest of the CC risks.

Table 3.4 Classification of global climate change risks on cities and buildings

Classification of CC risks on cities	Africa	Asia	Europe	North America	Latin America	Total
Future positive impact	1	1	1	1	1	1
Negative impact already occurred	15	13	11	13	13	65
Future negative impacts	2	4	6	5	4	21
Total of CC negative impacts	17	17	17	18	17	86

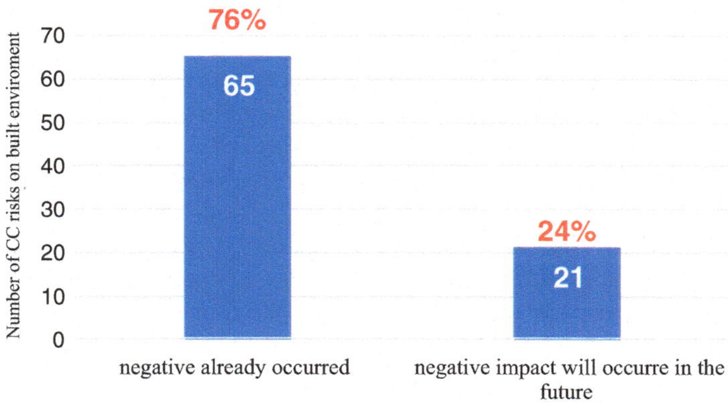

Fig. 3.17 Classification of climate change impacts: frequency and severity gradually increase over time. (Source: Developed by authors)

In this book, focuses will be on developing countries such as these in Africa, particularly Egypt.

A question that needs to be addressed is that what is the importance of adaptation, if the world mitigates GHG emissions?

According to the Intergovernmental Panel on Climate Change (IPCC), if emissions were completely halted, even though this is unexpected in the most optimistic scenarios, climate change severity would continue to increase.

Thus, adaptation to climate change is inevitable; therefore, Chap. 4 will address climate change adaptation measures globally to offset climate change risks.

3.3 Conclusion

This chapter mainly investigates the consequences of climate change risks on the built environment and infrastructure in cities, including agriculture and food security, ecosystems, forests, water, health, coastal and flood-prone zones, tourism, energy and economy risks related to the built environment that is known as climate change (CC) indirect impact. Also, the chapter clarifies the direct damage from CC risks while addressing the direct and indirect impacts on the built environment; thus the whole image about climate change consequences was completed including pros and cons.

In addition, the chapter presents weather-related disasters around the world in 5 years between 2010 and 2015 to emphasize such significance. Moreover, it discusses which climate change risks happened and what will happen in the future. Furthermore, the classification of continents' exposure is carried out.

The investigation indicates that the CC impacts on cities and buildings are negative. About 75% of these risks on cities and buildings have already occurred; thus risks' frequency and severity gradually increase over the time. The investigation also points out that Africa is the most effected continent, but in the future risks will extend to affect all continents equally which demonstrate that CC risks on cities and buildings are global risk directly affecting all mankind.

All of the previous risks reflect on urban areas and city in many forms. Migration of farmers to urban areas due to droughts will increase the stress on receiving cities. The increase in ecological migration isolated by roads and settlements will, in return, increase car accidents. Landslides will destroy roads and buildings near mountains, settlements near forests face the risk of forest fire and water stress decreases water access. Soil salinization damages foundations of buildings and roads. Heat stress decreases productivity of construction workers. Furthermore, an increase in temperatures increases energy demand and affects air quality in cities.

Thus, the importance of reducing emissions arises; however, according to the Intergovernmental Panel on Climate Change (IPCC), if emissions were completely halted, even though this is unexpected in the most optimistic scenarios, CC severity would continue to increase; therefore CC adaptation is inevitable.

References

1. Elisabeth Meze-Hausken, "Migration caused by climate change: how vulnerable are people inn dryland areas" Springer, vol. 5, no. 4, pp. 379–406, December 2000.
2. Oli Brown, "Migration and Climate Change No. 31" International Organization for Migration Geneva, Geneva, 2008.
3. UNEP, "AFRICA Atlas of Our Changing Environment" United Nations Environment Programme, 2008.
4. Nicole Roskos, "Caution: Wildlife Crossin" The Wildlife Watch Binocular, 2004. [Online]. Available at: http://www.wildwatch.org/Binocular/bino01/caution.html. (Accessed: June 26, 2015).
5. Teo Kermeliotis and Aja Harris, "elephant cross the road threat human's safety" CNN, 17 January 2014. [Online]. Available at: http://edition.cnn.com/2014/01/17/world/africa/why-did-elephant-cross-road-drunks/. (Accessed: March 16, 2016).
6. Beniston, UBA, "The 2003 heat wave in Europe: a shape of things to come?" European Environment Agency (EEA), 2005.
7. Kevin Schaefer et al, "Policy Implications of Warming Permafrost" United Nations Environment Programme, 2012.
8. "Landslide Questions" [Online]. Available at: http://www.weatherwizkids.com/?page_id=1326. (Accessed: June 26, 2015).

9. Charles Harris, "Recent Warming of European Permafrost: Evidence from Borehole Monitoring" *ScienceDirect,* vol. 92, no. 3–4, p. 117–171, 2009.
10. Mike Flannigan, "Impacts of climate change on fire activity and fire management in the circumboreal forest" *Global Change Biology,* pp. 549–560, 2008.
11. European Commission Joint Research Centre Institute for Environment and Sustainability, "Forest Fires in Europe, Middle East and North Africa" Directorate-General Environment, European Commission, 2014.
12. Union of Concerned Scientists, "Is Global Warming Fueling Increased Wildfire Risks?" Union of Concerned Scientists, [Online]. Available at: www.ucsusa.org/global_warming/science_and_impacts/impacts/global-warming-and-wildfire.html. (Accessed: June 22, 2015).
13. "Greece wildfire kills at least 50 near Athens, residents flee homes" hindustantimes, July 18, 2018. [Online]. Available at: www.hindustantimes.com/world-news/greece-wildfire-kills-at-least-20-near-athens-residents-flee-homes/story-HS4uw9neixyibDZlqck1gL.html. (Accessed: August 20, 2018).
14. "International Decade for Action 'Water for Life' 2005–2015" UN, [Online]. Available at: http://www.un.org/waterforlifedecade/asia.shtml. (Accessed: March 17, 2016).
15. "Water and cities" UN, November 24, 2014. [Online]. Available at: http://www.un.org/waterforlifedecade/water_cities.shtml. (Accessed: June 22, 2015).
16. Carmen Revenga, "Access to renewable water in 2025" World Resources Institute, 2000.
17. Katalina Engel et al, "Big Cities big Water big Challenges Water in an Urbanizing World" WWF, Berlin, Germany, 2011.
18. Kerstin K. Zander et al, "Heat stress causes substantial labour productivity loss in Australia" *Nature Climate Change Letter,* 2015.
19. Karin Lundgren, Kalev Kuklane and Vidhya Venugopal, "Occupational heat stress and associated productivity loss estimation a case study from workplaces in Chennai, India" Global Health Network, 2014.
20. Nadal, N et al, "Building Damage due to Riverine and Coastal Floods" *Journal of Water Resources Planning and Management,* vol. 136, no. 3, 2010.
21. The State of Queensland (Office of the Queensland Chief Scientist), "What are the consequences of floods?" [Online]. Available at: http://www.chiefscientist.qld.gov.au/publications/understanding-floods/flood-consequences. (Accessed: June 26, 2015).
22. "New York Hurricane Insurance: Fact File" Insurance Information Institute, [Online]. Available at: http://www.iii.org/article/new-york-hurricane-insurance-fact-file. (Accessed: March 18, 2016).
23. C. P. Kumar, "Climate Change and Its Impact on Groundwater Resources Research Inventy" *International Journal of Engineering and Science,* 2012.
24. "Salinity Intrusion in a Changing Climate Scenario will Hit Coastal Bangladesh Hard" The World Bank Group, February 17, 2015. [Online]. Available at: http://www.worldbank.org/en/news/feature/2015/02/17/salinity-intrusion-in-changing-climate-scenario-will-hit-coastal-bangladesh-hard. (Accessed: August 20, 2018).
25. "Disaster Summary Sheet Tropical Cyclones", ACAPS the Assessment Capacities Project, July 2011.
26. "Climate Change & Tourism, World Tourism Organization Network" [Online]. Available at: http://sdt.unwto.org/en/content/climate-change-tourism. (Accessed: June 26, 2015).
27. "FAQ - Climate Change and Tourism, World Tourism Organization Network" [Online]. Available at: http://sdt.unwto.org/content/faq-climate-change-and-tourism. (Accessed: June 26, 2015).
28. J.C.R. Hunt et al, "Introduction. Climate change and urban areas: research dialogue in a policy framework" *The Royal Society,* 2007.
29. Sheng-chieh Chang, "Energy Use" Environmental Energies Technology Division, June 23, 2000. [Online]. Available at: https://web.archive.org/web/20090311050754/http://eetd.lbl.gov/heatisland/EnergyUse/. (Accessed: August 15, 2015).
30. M Santamouris et al, "The impact of urban climate on the energy consumption of buildings" *Elsevier,* vol. 70, no. 3, pp. 201–216, 2001.

31. United States Environmental Protection Agency, Office of Atmospheric Programs, "Climate Change in the United States: Benefits of Global Action" United States Environmental Protection Agency, 2015.
32. Bernhard Lehnera, Gregor Czischb and Sara Vassoloa, "The impact of global change on the hydropower potential of Europe: a model-based analysis" *Elsevier,* December 19, 2003.
33. H. (Bert) W. Whittington, Gareth P Harrison, "Susceptibility of the Batoka Gorge hydroelectric scheme to climate change" *Elsevier,* vol. 264, no. 1–4, pp. 230–241, 2002.
34. Kristin Linnerud, Torben K. Mideksa and Gunnar S. Eskeland, "The Impact of Climate Change on Nuclear Power Supply" *CAMBRIDGE MA THEN CLEVELAND-OH,* January 2011.
35. Bernard Aebischer, Giacomo Catenazzi and Martin Jakob, "Impact of climate change on thermal comfort, heating and cooling energy demand in Europe" ECEEE, 2007.
36. Ulisses Confalonier et al, "Human Health". In: Climate Change 2007: Impacts, Adaptation and Vulnerability Cambridge University Press, Cambridge, UK, 2007.
37. Peter Feiden, "Adapting to Climate Change:Cities and the Urban Poor" International Housing Coalition, August 2011.
38. Twenty First Session of the Governing Council, "Sustainable Urbanization: local action for urban poverty reduction, emphasis on finance and planning" in *Sustainable Urbanization,* Nairobi, Kenya, 16–20 April 2007.
39. the Urban Land Institute, "What the Real Estate Industry Needs to Know about the Insurance Industry and Climate Change" Urban Resilience Program, 2014.
40. The Institutional Investors Group on Climate Change (IIGCC), "Managing the climate risks and opportunities in your real estate investments" (IIGCC), January 21, 2015.
41. Sven Bienert, "Extreme weather events and property values" Urban land institute, 2014.
42. Akbari H, " Energy Saving Potentials and Air Quality Benefits of Urban Heat Island Mitigation" Lawrence Berkeley National Laboratory, 2005.
43. Center for Disease Control and Prevention, "Extreme Heat: A Prevention Guide to Promote Your Personal Health and Safety" 2006.
44. James W, "Green roads: research into permeable pavers" 2002.
45. Jon Barnett and Michael Webber, "Accommodating Migration to Promote Adaptation to Climate Change" The World Bank Development Economics Office of the Senior Vice President and Chief Economist, 2010.
46. Refugee Studies Centre for the Conflict, Humanitarian and Security Department et al, "Environmentally Displaced People: Understanding The Linkages Between Environmental Change, Livelihoods and Forced Migration a Policy Briefing" December 20, 2007.
47. Michelle Yonetani, "Global Estimates 2014 People displaced by disasters" Internal Displacement Monitoring Centre, September 2014.
48. Alexandra Fielden Intern, "NEW ISSUES IN REFUGEE RESEARCH Research Paper No. 161 Ignored Displaced Persons: the plight of IDPs in urban areas" Policy Development and Evaluation Service, UNHCR, July 2008.
49. Nansen Initiative Secretariat International Environment House, "Linking Human Mobility, Disasters and Disaster Risk Reduction" Switzerland, September 2014.
50. Rebecca Anne Dixon and Ambassador Teresita Schaffer, "Pakistan Floods: Internally Displaced People and the Human Impact" Center for Strategic and International Studies -CSIS, November 2010.
51. United Nations, "Next Steps: New dynamics of displacement" [Online]. Available at: www.un.org/en/globalissues/briefingpapers/refugees/nextsteps.html. (Accessed: July 26, 2015).
52. Christine Mai-Duc, Veronica Rocha And Joseph Serna, "California storm: 113,000 still without power, travel disrupted" Los Anglos times California, December 11, 2014.
53. BBC, [Online]. Available at: Dorset Police said the high winds had brought down dozens of trees, causing numerous roads to be blocked. [Accessed: January 15, 2015].
54. Michelle Davis and Steve Clemmer, "Power Failure" Union of concerned scientists, 2014.
55. Climate Central, "Weather-Related Blackouts Doubled Since 2003" Climate Central, 2014.

56. "Hurricane Irma: Two-thirds of Florida without power" BBC News, 12 September 2017. [Online]. Available at: http://www.bbc.com/news/world-us-canada-41234292. (Accessed: December 2, 2017).
57. Torbay Council, "Extreme weather resilience report" Torbay Council, Torbay, October 2014.
58. "Hurricane Katrina hit New Orleans 10 years ago, today: Here's how it happened" TheJournal.ie, [Online]. Available at: http://www.thejournal.ie/hurricane-katrina-in-real-time-events-2290072-Aug2015/. (Accessed: March 18, 2016).
59. Rosie Manins, "Hurricane relief for Puerto Rico, Virgin Islands" Cross Roads News, November 17, 2017. [Online]. Available at: http://www.crossroadsnews.com/news/hurricane-relief-for-puerto-rico-virgin-islands/article_51db3278-c411-11e7-b17a-87ef797f8b4d.html. (Accessed: December 17, 2017).
60. "Western Indian State Tries to Cope With Record Heat Wave" Wall Street Journal, May 21, 2016. [Online]. Available at: www.wsj.com/articles/western-indian-state-tries-to-cope-with-record-heat-wave-1463846290. (Accessed: December 3, 2017).
61. "Downpours flood parts of French Riviera; at least 20 dead" CNN, October 5, 2015. [Online]. Available at: http://edition.cnn.com/2015/10/04/europe/france-french-riviera-flooding/. (Accessed: March 26, 2016).
62. "Floods, earthquakes, wildfires and heat waves: the worst natural disasters of 2015" CBC/Radio-Canada, December 2015. [Online]. Available at: http://www.cbc.ca/news/world/year-end-2015-natural-disasters-1.3346639. (Accessed: March 20, 2016).
63. "Deadly flash floods hit Egypt's Alexandria" Al Jazeera Media Network, Oct 2015. [Online]. Available at: http://www.aljazeera.com/news/2015/10/deadly-flash-floods-hit-egypt-alexandria-151026085943090.html. (Accessed: March 20, 2016).
64. "Natural catastrophes and man-made disasters in 2014" Swiss Re Ltd, Switzerland, 2015.
65. Aon Benfield, "Annual Global Climate and Catastrophe Report" Impact Forecasting, 2012.
66. "Natural catastrophes and man-made disasters in 2013" Swiss Re, 2014.
67. CLIMATE GUEST CONTRIBUTOR, "Top 10 Global Weather Events of 2011" 2 JAN 2012. [Online]. Available at: http://thinkprogress.org/climate/2012/01/02/396307/top-10-global-weather-events-of-2011/. (Accessed: March 20, 2016).
68. "World Disasters 2010 Timeline" World News Atlas, February 9, 2010. [Online]. Available at: http://www.mapreport.com/years/2010/d.html. (Accessed: March 21, 2016).
69. "Rain from Typhoon Megi" The Earth Observatory, NASA Goddard Space Flight Center, [Online]. Available at: http://earthobservatory.nasa.gov/NaturalHazards/view.php?id=46517. (Accessed: March 20, 2016).
70. "A tragedy in Oaxaca" The Economist Newspaper, September 28, 2010. [Online]. Available at: http://www.economist.com/blogs/americasview/2010/09/floods_mexico. (Accessed: March 20, 2016).
71. Brent Sohngen, Robert Mendelsohn, and Roger Sedjo, "A Global Model of Climate Change Impacts on Timber Markets" *Journal of Agricultural and Resource Economics,* 2001.
72. Climate Impacts Group,Washington University, "IMPACTS OF CLIMATE CHANGE ON PNW TIMBER PRODUCTION" Washington University, June 2004.
73. Roger Sedjo, "The Future of Trees, climate change and the timber industry" Resources Publications, 2010.
74. "Desertification Land Degradation &Drought (Dldd) - Some Global Facts &Figures" [Online]. Available at: www.Un.Org/En/Events/Desertificationday/Background.Shtml. (Accessed: May 24, 2015).
75. Tierney Smith, "Climate change, desertification and migration" [Online]. Available at: www.rtcc.org/2012/04/27/climate-change-desertification-and-migration-connecting-the-dots/. (Accessed: May 25, 2015).
76. Michelle Leighton Schwartz et al, "Desertification and Migration: Mexico and the United States" Natural Heritage Institute, San Francisco, California, 1994.
77. "Desertification Increases Poverty and Contributes to Migration in Latin America and the Caribbean" June 18, 2007. [Online]. Available at: www.cepal.org/cgi-bin/getProd.asp?xml=/

prensa/noticias/comunicados/0/28890/P28890.xml&xsl=/prensa/tpl-i/p6f.xsl&base=/prensa/tpl/top-bottom.xslt. (Accessed: May 22, 2015).
78. "Species and ecosystem impacts" David Suzuki Foundation, [Online]. Available at: www.davidsuzuki.org/issues/climate-change/science/impacts/species-and-ecosystems/. (Accessed: May 15, 2015).
79. Renee Cho, "Climate Change Poses Challenges to Plants and Animals" Earth Institute, Columbia University, February 3, 2015.
80. Jim Robbins, "Can Ecological Corridors Heal Fragmented Landscapes" Yale Environment 360, October 2011. [Online]. Available at: e360.yale.edu/feature/ecological_corridors_connecting_fragmented_pockets_of_wildlife_habitat/2450/. (Accessed: October 12, 2015).
81. Marie Bonnin, The Pan-European Ecological Network, Council of Europe, 2007.
82. Sameer Karki, "Community Involvement in and Management of Forest Fires in South East Asia" Project FireFight South East Asia, 2002.
83. "PREVENTION AND CONTROL OF FOREST FIRES" in *Latin American And Caribbean Forestry Commission Item 7(a) of the Provisional Agenda,Twenty-First Session*, Santa Fe de Bogota, Colombia, 4–8 September 2000.
84. GREENTUMBLE EDITORIAL TEAM, "Water Scarcity in Africa: Issues and Challenges" Greentumble, 20 SEPTEMBER 2015. [Online]. Available at: https://greentumble.com/water-scarcity-in-africa-issues-and-challenges/. (Accessed: January 24, 2016).
85. The Intergovernmental Panel on Climate Change (IPCC), "Climate Change 2007: Impacts, Adaptation and Vulnerability" Cambridge University Press, 2007.
86. Uma Langkulsen, Nuntavarn Vichit-Vadakan, and Sasitorn Taptagaporn, "Health impact of climate change on occupational health and productivity in Thailand" *Glob Health Action,* 2010.
87. Tord Kjellstrom et al, "The Direct Impact of Climate Change on Regional Labor Productivity" *Archives of Environmental and Occupational Health,* pp. 217–227, August 7, 2010.
88. Noah Kofi, "Flooding And Physical Planning In Urban Areas In West Africa: Situational Analysis Of Accra, Ghana" Heriot Watt University, United Kingdom, 2009.
89. Kourtnii S. Brown, "Flooding in Asia's Megacities", The Asia Foundation, 2012.
90. Robert C. Brears, "Preparing for floods: five lessons from Europe", [Online]. Available at: www.thethirdpole.net/preparing-for-floods-five-lessons-from-europe/. (Accessed: July 27, 2015).
91. "More Extreme Weather and the U.S. Energy Infrastructure", National Wild Life Federation, 2011.
92. Patricia Romero Lankao, "The Challenge of Climate Change in Latin America" London School of Economics and Political Science, December 2008.
93. Claire Bedelian, "Ecosystem management, Environmental governance and Climate change", December 2013. [Online]. Available at: http://na.unep.net/geas/getUNEPPageWithArticleIDScript.php?article_id=107. (Accessed: June 26, 2015).
94. "The Economics of Climate Change in Southeast Asia", Asian Development Bank, 2009.
95. "East Riding coastal erosion", June 24, 2014. [Online]. Available at: www.theguardian.com/environment/gallery/2014/jun/24/east-riding-coastal-erosion-in-pictures. (Accessed: July 14, 2015).
96. "US West Coast Erosion Spiked In Winter 2009, Previewing Likely Future As Climate Changes", U.S. Department of the Interior, U.S. Geological Survey, December 7, 2011. [Online]. Available at: www.usgs.gov/newsroom/article.asp?ID=2849from=rss_home. (Accessed: July 18, 2016).
97. Nelson Guillermo et al, "Coastal erosion along the Caribbean coast of Colombia: Magnitudes, causes and management", *Elsevier,* vol. 114, 2015.
98. Tej Vir Singh, "Critical Debates in Tourism", Channel View Publications, 2012.
99. Elizabeth Burakowski and Matthew Magnusson, "Climate Impacts on the Winter Tourism Economy in the United States" Natural Resources Defense Council, 2012.

100. The Intergovernmental Panel on Climate Change (IPCC), Climate Change 2001: Impacts, Adaptation and Vulnerability. [Online]. Available at: http://www.grida.no/publications/other/ipcc_tar/. (Accessed: 18 April 2015).
101. The Intergovernmental Panel on Climate Change (IPCC), "Climate Change 2014: Impacts, Adaptation and Vulnerability" Cambridge University Press, Cambridge, UK, 2014.
102. Shushi Peng et al, "Surface Urban Heat Island Across 419 Global Big Cities", *Environmental Science & Technology,* pp. 696–703, 2011.
103. Diego J. Rodriguez, "Interlinkages and trade-offs between water and energy", The World Bank, 2014.
104. Van Butsic, "Climate Change and Housing Prices: Hedonic Estimates for Ski Resorts in Western North America", Federal Reserve Bank of San Francisco, November 2009.
105. "Real Estate 2020 Building the future", PwC Global Real Estate Leadership Team, 2014.
106. "Severe flooding hits southeast Asia", 27th December 2014. [Online]. Available at: www.aljazeera.com/news/asia-pacific/2014/12/severe-flooding-hits-southeast-asia-201412275647819859.html. (Accessed: July 28, 2015).
107. "Disasters and displacement in a changing climate", Centre Oxford Department of International Development, University of Oxford, May 2015.

Chapter 4
Global Climate Change Adaptation Measures in Cities and Urban Areas

Abstract Climate change (CC) is inevitable! The need for urgent response to climate change impacts is momentarily taking place to mitigate such impacts of severe events on the most vulnerability sectors, whether on current risks or future risks. Nonetheless, these responses are not at the same momentum with the severity of CC risks. Throughout history, civilizations have been associated with climate. People and societies have adjusted and coped with climate at varying degrees of success. Drought, for example, has been at least partly responsible for the rise and fall of civilizations. Scientifically, the idea of adaptation emerged in 1991 by the US National Academy of Sciences. Climate change adaptation (CCA) measures reduce the adverse effects of climate change risks by taking appropriate actions to prevent or minimize the damage resulting in exposure to hazards or taking advantage of opportunities that increase resilience to climate change and sustainability. This chapter identifies adaptation measures in cities, including urban and rural areas that aid in preventing, moderating or adapting to the consequences of CC risks. These measures are discussed specifically on ten sectors (flood prone areas, urban areas, infrastructure and activities in the city such as fresh water, energy, public health, agriculture, forest, biodiversity and economy). Also, the share of Africa, Asia, Europe, North America and Latin America from these measures will be identified. Finally, the chapter explains the role and significance of CCA measures in enhancing and/or upgrading cities and buildings' retrofitting.

4.1 Global Climate Change Adaptation Measures

The following sections discuss climate change adaptation (CCA) measures adopted in Africa, Asia and Europe, as well as in North America and Latin America, particularly coastal and flood-prone areas, urban areas and infrastructure, fresh water as well as energy, public health, agriculture and forest and biodiversity and economy to show how to cope with climate change severe events.

M. M. Aboulnaga et al., *Urban Climate Change Adaptation in Developing Countries*, https://doi.org/10.1007/978-3-030-05405-2_4

4.1.1 Coastal Areas and Flood-Prone Measures

Extreme weather events have been increasing since 1950 due to climate change impacts, mainly floods, forest fires and extreme heat and cold [1]. Early warning systems' phases include detection, analysis, forecasting and warning and then response of decision-makers [2]. Moving properties from areas that are likely to be flooded and the resettlement of vulnerable communities should be a strategy of last option. Resettled people are exposed to risk of landlessness, unemployment, food insecurity, increased morbidity and community disarticulation in host communities. Hence, it should be linked with sustainable resource management approach to avoid negative impacts of resettlement [3].

Also, object protection adaption (OPA) measures consist of building walls around properties, using waterproof low parts of houses and raising houses on stilts [4]. Dry flood-proofing measures encompass waterproof coating for exterior walls, movable barrier to seal opening and elevating opening above flood level. Therefore, these measures could be applied to coastal cities that are vulnerable to CC risks such as New York and Miami, United States; Tokyo, Japan; Rotterdam, The Netherlands; Shanghai, China; Kolkata, India; and Alexandria, Egypt.

Other measures include wet flood-proofing measures such as elevating buildings above flood level as depicted in Fig. 4.1 or allowing flood water to enter enclosed area such as basement throw openings (Fig. 4.2). Sea-level rise (SLR) adaptation strategies are divided into three sub-strategies: (a) protection, (b) accommodation and (c) retreat. The protection strategy is divided into two types: (i) hard engineering measures such as building a sea wall and groyne and boulder barriers [5] and (ii) soft engineering measures, for instance, beach nourishment and managed retreat multipurpose artificial coral reefs and beach nourishment by sand and green belts [6]. The accommodation strategy encompasses raising the ground level or improving drainage facilities, restoring sand dunes and improving warning systems. The retreat strategy depends on relocating developments further from low lying areas [7] as illustrated in Table 4.1 (below).

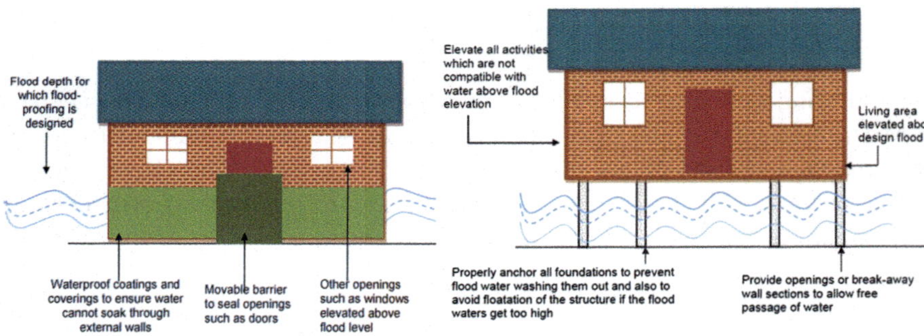

Fig. 4.1 Dry (left) and wet (right) flood-proofing measures. (Source: Linham and Nicholls, 2010)

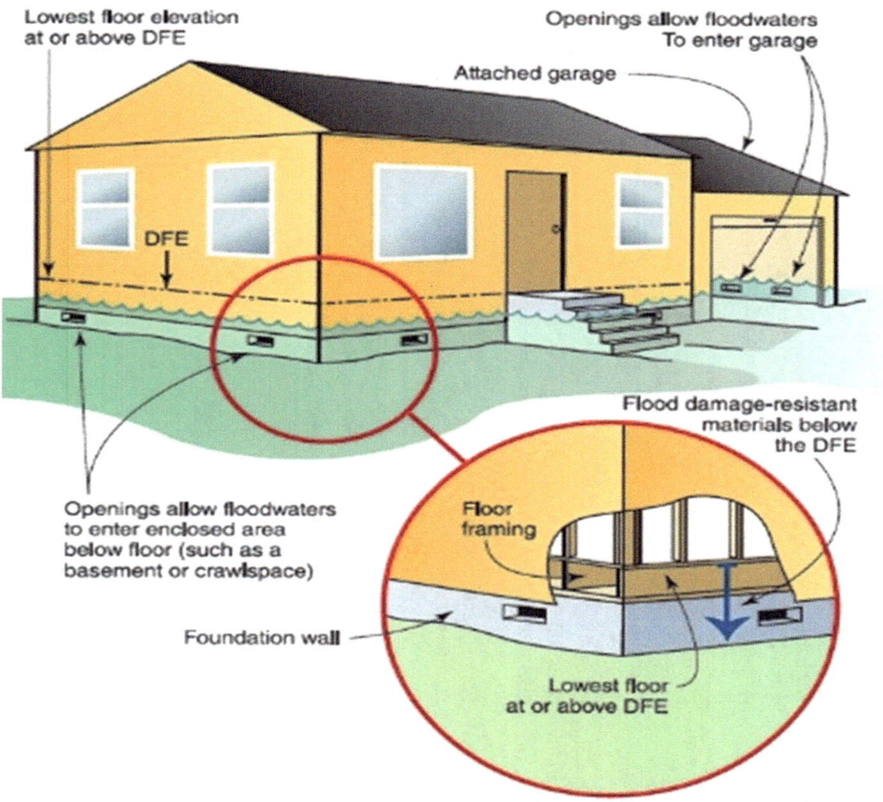

Fig. 4.2 Wet flood proofing strategies. (Source: Federal Emergency Management Agency)

Climate change adaptation (CCA) through land use planning will minimize its risks [8]. Decision-makers and planners can take specific actions in land use planning to adapt to the impacts of CC, for instance, land zoning according to the vulnerability to climate change, imposing special regulations to risk areas and imposing use on restrictions on flood areas for 100 years, particularly utilizing the areas in low-density activities, e.g. agricultural or recreational uses, and taking measures to adapt to CC [9]. Adaptation costs will be insignificant in comparison with losses due to severe events and associated risks and damages [10].

Adaptation of households to climate change through awareness raising is vital, and it could be performed by informing stakeholders about the vulnerability, type of anticipated risks and the measures to be taken to adapt to CC. Awareness raising enhances the adaptive capacity to lessen overall vulnerability [11]. Hence, it is an important measure, especially in rural areas due to the fact that such community is accustomed to observe CC impacts without taking action [12].

Table 4.1 Sea-level rise adaptation strategies

Strategies		Features	Methods	Site advantages	Diagrams
Protection (defence)	Hard	Dikes Levees Sea Walls Groins	Build physical barriers to block water	For hard-to-move facilities and infrastructure on flat ground	
	Soft	Mangroves Wetlands Sand dunes Tidal flats	Create buffer with vegetation or landforms	For maintaining shoreline at sites with existing coastal forest or sand dunes	
Accommodation		Raising level Desalination Drainage Alarm system	Upgrade functions while maintaining location	For redevelopment projects or facilities at sites without high ground nearby	
Retreat		Relocation Abandonment	Relocate facilities to low-risk uplands	For residential and public facilities at sites with low-risk uplands nearby	
Attack		Land reclamation Piers Ports Harbors	Extend facilities towards water	For facilities requiring direct access to water	

Yumi Lee [7]

Also, preserving the protracted areas that inhabit mangroves (part of the ecosystems) through coastal planning and management, can be utilised in eliminating stresses and reducing the adverse effects on mangrove ecosystems, hence contributes to resilience towards climate change effects [13].

4.1.2 Urban Areas' Measures

Green infrastructure in cities can play a major role in mitigating climate change. This includes parks, gardens, wetlands and green roofs (Fig. 4.3). Green infrastructure improves the environment and acts as an essential tool for climate change mitigation (CCM), since it absorbs carbon dioxide (CO_2) and pollutants, therefore enhancing the overall air quality as well as city liveability [14]. It also assists in lessening the CC risks and aids in the adaptation to CC as it filters underground water from pollution, conserves ecosystem and enriches the capacity of soils for water retention in flood-prone areas as well as enhances outdoor thermal mass and reduces urban heat island (UHI) effect [15]. In addition, buildings' regulations are used to set the requirements in cities and buildings to integrate CC impacts and measures into the design process in order to strengthen future adaptation responses to CC risks. Moreover, measures such as retrofitting, flood-proof structures, green building codes and land use policies play a significant role in preserving the ecosystem [16].

Fig. 4.3 A type of urban agriculture in cities. (Source: University of Arkansas Community Design Centre)

Selective relocation of property and infrastructure to non-flood-prone areas is also possible. The land may be utilized for recreation and agriculture or as habitat to preserve ecosystem as shown in Fig. 4.4. Nonetheless, it is a long-term solution to sea-level rise, but it needs community awareness that can be achieved through workshops, seminars and use of the media and others [17]. However, it is an expensive solution due to several reasons, among them financial compensation of dwellers, the cost of relocating infrastructure and loss of real estate investment as well as the generation of new social risks to community [18].

Waste management is an important measure in CCM and CCA. It is achieved through waste minimization in order to reduce the use of energy and raw materials. Also, it can be done through reusing, recycling and converting waste to energy or by the least prioritized action, which is disposal of wastes (Fig. 4.5). If landfills inundated due decomposition of floods materials, this will lead to rodent infestations, disease outbreak and groundwater contamination [19]. Adaptation by minimizing accumulation of waste, recycling organic wastes and relocating waste routes away from surface water supplies or flood plains increases the efficiency of transfer and treatment of waste [20].

Sustainable urban development (SUD) as a part of CCA strategy enhances the sustainability of cities in terms of population's health, comfort, safety and life qual-

Fig. 4.4 Retreat strategy. (Source: Mayor Michael R. Bloomberg, 2013)

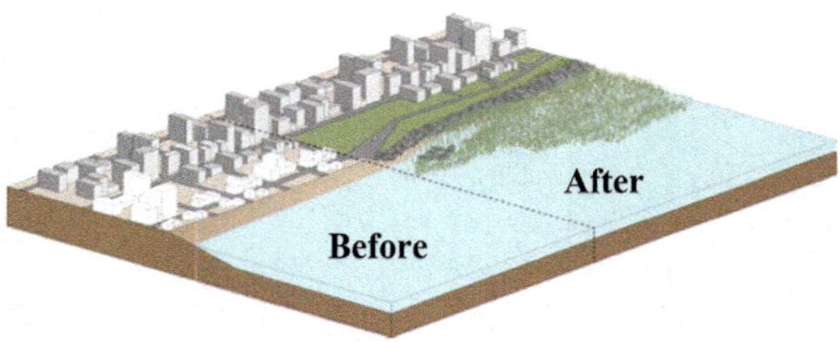

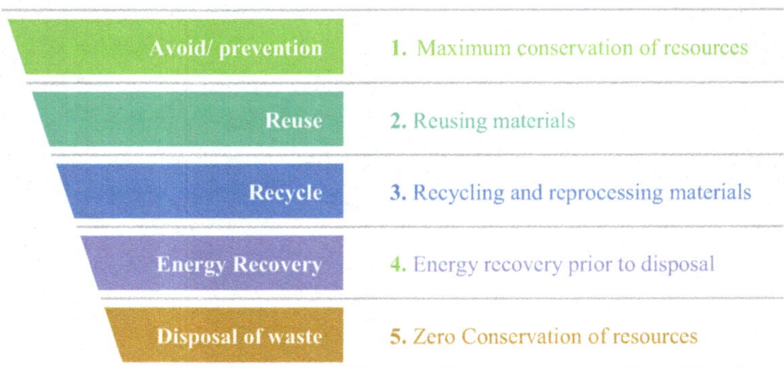

Fig. 4.5 Waste minimization hierarchy. (Re-illustrated after greenhome.ie - Source: https://www.greenhome.ie/Waste/Waste-Preventionimage)

ity (livability) and encourages sustainability of investments in utilities and infrastructures; this is a mere response to climate change. Sustainable urban development will also provide better planning, accurate prediction and bold decision support as well as a training tool based on information of present and future climate change risks that should be used in supporting and feeding in urban plans [21]. Positive impacts of SUD are buttressing food security, saving community, strengthening institutional systems and building the resilience of communities to face stresses of CC as well as enhancing the sustainability of natural resources [22]. Figure 4.6 (above) presents sustainable cities' pillars that combine between saving natural resources, providing standard of living and wellbeing and enhancing green economic growth.

Early warning systems predict and forecast climate disasters and warn vulnerable communities and then monitor hazard parameters (Fig. 4.7). Systems include detection from sensors which send data to the control centre that analyse the data and trigger the alarm in case of floods. It is essential to generate accurate warnings in a timely fashion. Early warning is a major element of disaster risk reduction because it prevents loss of life and reduces the economic losses and impact of disasters and helps stakeholders make a decision and enhances the resilience of communities to adapt to climate change hazards [23]. Vulnerability mapping also supports

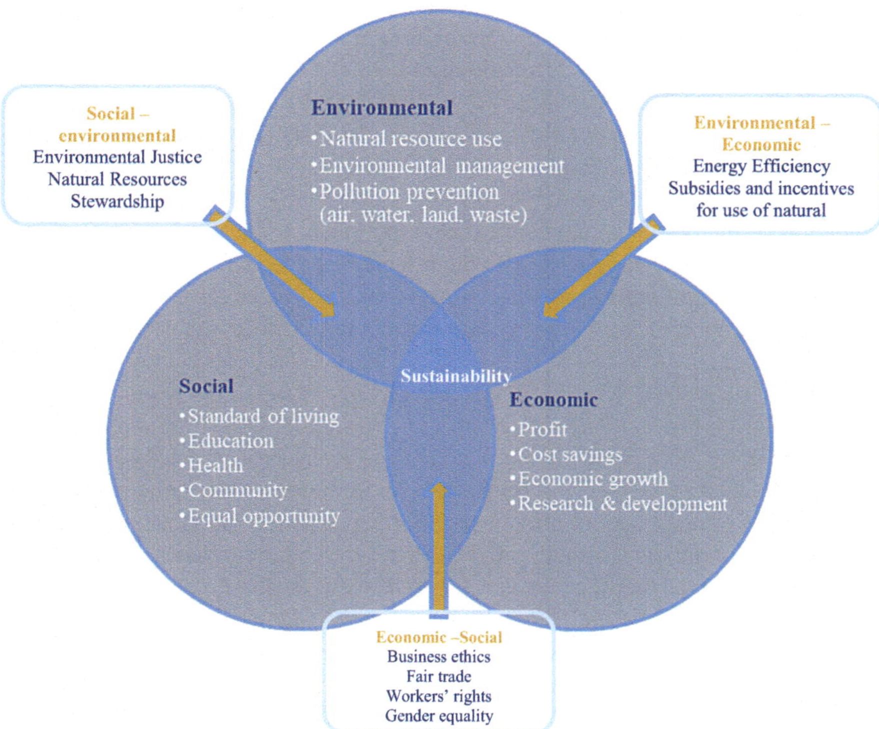

Fig. 4.6 Sustainability pillars and dimensions. (Re-illustrated after University of Michigan, 2002)

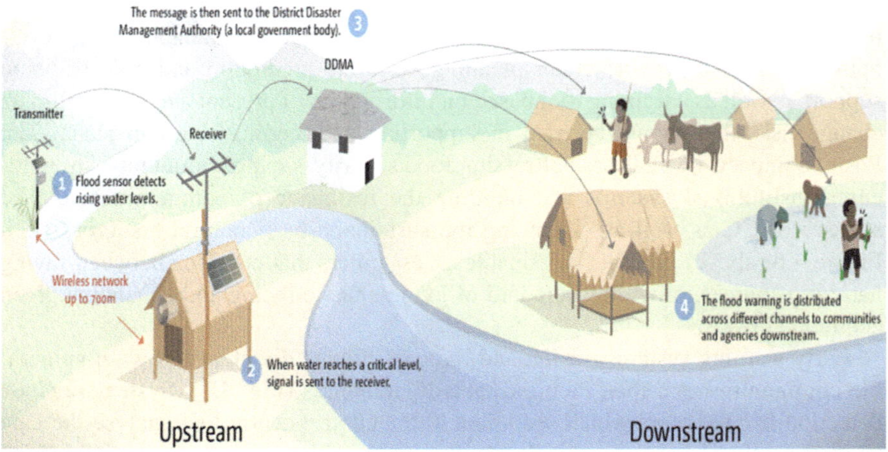

Fig. 4.7 Community-based flood early-warning system. (Source: UN Framework Convention on Climate Change, 2014)

the assessment of risk, motivates people to prepare for disaster prevention, and responds in vulnerable areas as well as determines the level of sensitivity of the areas exposed to climate change impacts [24].

City climate monitoring systems are used to monitor and forecast weather in the city by satellite observations, ground-based data and forecast models. The system then analyses data to identify mean values, trends and variations of climate to provide statistical analysis about climate variations in temperatures, humidity and precipitation and then create data base containing records of measurements over time to evaluate current climate and predict future climate [25]. This system helps stakeholders take the best adaptations measures and enhance capability to prepare and respond to climate change hazards, which increases probability of recovery from risk with minimum damage to society, economy and the environment. Figure 4.8 illustrates the monitoring network station mounted on a lamppost in one of the streets in Serbia.

Urban agriculture is growing plants and raising animals in or around a village, town, city or on top roofs as shown in Fig. 4.9 (below). Urban agriculture can be accompanied by other complementary activities such as processing and distributing food and collecting and reusing food waste and rainwater. Also, it is important to raise public awareness about urban agriculture such as the practice of agriculture, caring for crops and its advantages [234]. Urban agriculture reduces the impacts of floods and landslides in urban areas because it works as floodplain catchment, provides food sources and thereby reduces hunger and malnutrition, creates new green jobs and reduces the heat island effect by providing shade. Urban agriculture also assists in filtering underground water and air by absorbing pollution, reducing energy use for transport of food production and providing reuse of organic wastes as fertilizer [26].

Fig. 4.8 Monitoring network station in Serbia mounted on a lamppost. (Source: Ivan secerov, 2015)

Fig. 4.9 Roof top garden. (Source: Raeky, Creative Commons)

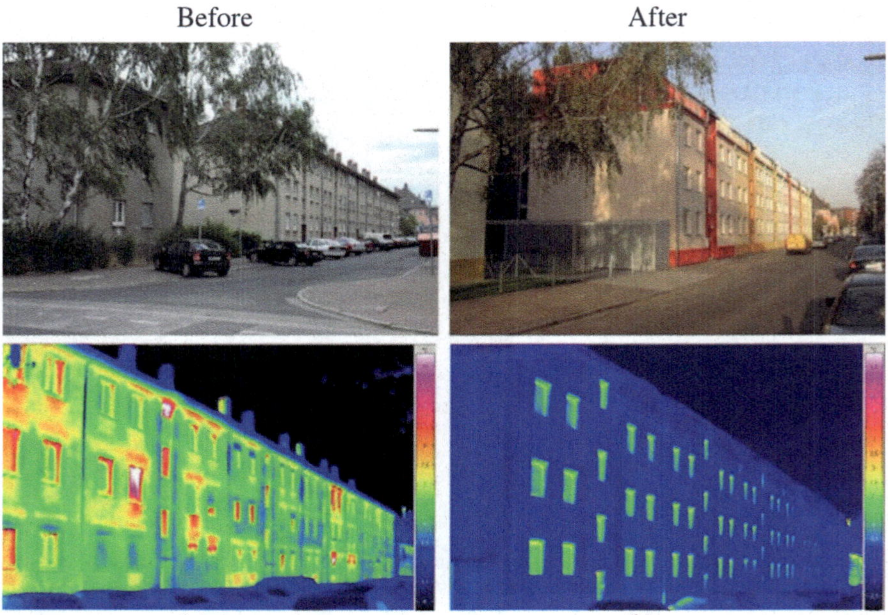

(a) (b)

Fig. 4.10 Thermal image of the building's envelop before and after incorporating insulations. (**a**) Envelop without insulation. (**b**) Envelop with insulation. (Source: International Passivhaus Association)

Cash transfer programmes are conditional upon the family's actions; hence, the government or a charity organization provides help for those who meet certain criteria [27]. Most informal urban settlements suffer from limited water availability, health care and poverty. Cash transfer programmes improve food security and access to health in addition to supporting adaptation and disaster risk reduction. This is provided by designing protection measures that strengthen adaptation opportunities to current and future CC impacts and face climate shocks and stresses in urban and rural areas [26].

Buildings' retrofitting affects and increases resilience to extreme weather events due to climate change such as heat waves, floods and air pollution, and their effects are mainly health impacts and injuries, urban heat island effect and inundation of buildings. Retrofitting functions occur by using passive cooling and flood-proofing systems. Passive cooling provides natural ventilation and cooling and enhances water supply and power security [28]. Also, passive cooling is a building design strategy to achieve comfort through non-mechanical cooling strategies to improve thermal comfort, including orientation, ventilation, windows, shading, insulation and thermal mass. Some of the passive-cooling principles are to choose materials with high thermal mass carefully to store coldness and prevent unwanted heat, to use double or triple glazing and shading windows and to maximize cross ventilation and buildings' envelop insulations to enhance energy efficiency [29]. In Fig. 4.10, the left thermal image shows the building's walls before insulation; it shows that the area coloured in yellow and red on the façade is storing a lot of heat. Whereas the right image shows the building after the insulation was added, the infra-red image turned blue and green indicating that the façade became cooler.

Flood-proofing strategies mean to reduce or avoid the impacts of flooding upon the buildings such as raising the buildings above flood level, choosing designs and buildings' materials which make structures more resilient to flood damage, preventing flood waters from infiltrating structures and other measures. Flood-proofing measures enhance development in flood hazard zone [30].

Retrofitting buildings' envelop reduces embodied and operating energy consumption. Energy saving is met by using strategies such as effective daylighting, solar water heating and others to adapt to climate change impact on energy. Retrofitting water efficiency is achieved by reducing water consumption and protecting water quality by using measures such as grey water recycling, upgrading irrigation conservation and fixing leaks as well as rainwater harvesting and others to adapt to droughts and increase water stress [31]. Climate change adaptation (CCA) strategies for buildings are divided into four types: (a) site strategies to increase site resilience to extreme weather events, (b) buildings' enclosure strategies to increase buildings' resilience to extreme weather events, (c) buildings' systems strategies to retrofit energy and water efficiency and (d) people strategies to educate residents of households.

Land use planning utilizes the land and its natural resources in an effective and efficient way through selecting the best land use options that assist in enhancing economic, social and environmental conditions of communities and safeguarding resources for the future, i.e. attaining sustainable development [32].

Climate change adaptation measures encompass land use planning tools such as official plans, zoning and development permits in flood-prone and waterfront for limiting its land uses on low density land activities (agricultural or recreational uses). For example, St. Joseph city in Michigan, USA, developed a future land use plan (Fig. 4.11) to minimize risks of floods [33]. Land use planning also adapts to CC risks through using some measures, such as environmental standards, energy and water efficiency, modified building codes and changes in urban form, to reduce dependencies on transportation, energy consumption and dealing with climate hazards. Some of CCA measures of coastal areas are mainly to prevent new permanent developments within risk areas or to develop soft engineering protections [34], which were explained in Sect. 4.1.1.

Urbanization occasionally causes habitat fragmentation, which increases the loss of biodiversity. Thus, urban planners must develop corridors for migration and enlarging core conservation areas to connect wildlife to face climate change impacts such as drought and increased temperature which will increase migration to find suitable life condition [35]. Figure 4.12 shows an ecological bridge connecting nature reserve in national parks in Singapore.

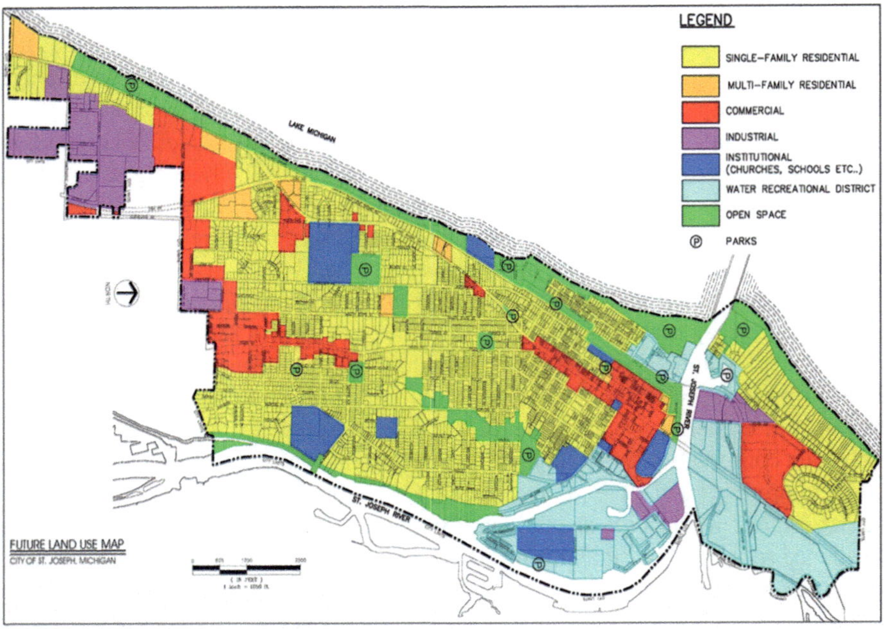

Fig. 4.11 Future land use plan of St. Joseph, Michigan city. (Source: City of St. Joseph)

Fig. 4.12 Ecological bridge connecting nature reserve in Singapore. (Source: Benjamin P, Creative commons)

4.1.3 *Infrastructure Measures*

There are several infrastructure measures that can be adopted to mitigate climate change risks, including: a) use of heat resistance materials on road surfaces and railway tracks to withstand higher temperatures; b) improve roads and strengthen bridges to be resilient to floods; c) upgrade trackside drains or use porous pavement (Fig. 4.13) to allow storm water movement through the surface to under-drain, thus accommodate with floods flow; and d) improve early warning system for roads and railways to reduce accident [36]. Measures also include sea walls and floods' defences to protect coastal roads and infrastructure against sea level rise (SLR) and from coastal erosion [37]. In addition, the relocation of infrastructure outside the risk zones in case of the adaptation cost is higher than the relocation cost; besides, the creation of marshlands and wetlands to act as buffer zone against SLR and severe floods [38]. It is important to note that railway systems are more vulnerable to climate change variability than the road system due to the fact that traffic can be redirected, but there are no alternative routes for rail networks [15].

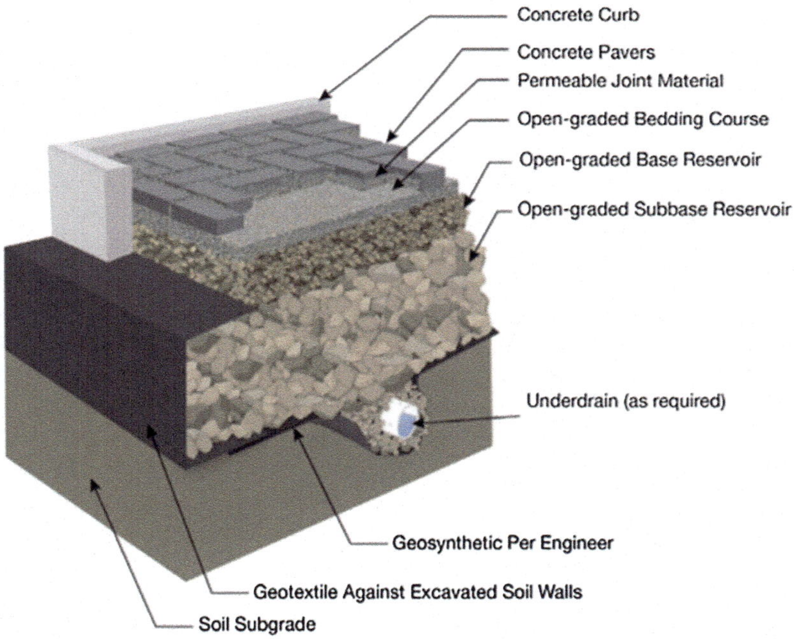

Fig. 4.13 Permeable pavement cross section. (Source: Interlocking Concrete Pavement Institute)

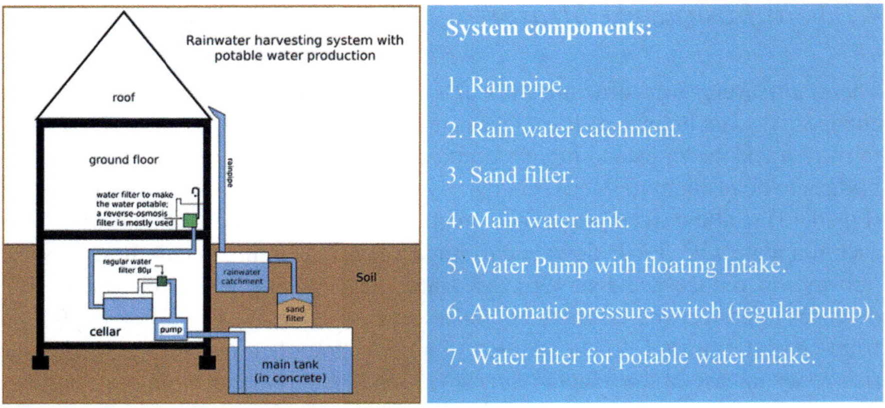

Fig. 4.14 Rainwater harvesting system in a low-rise residential building. (Source: https://de.wikipedia.org/wiki/Datei:Rainwater_harvesting_system.svg)

4.1.4 Fresh Water Measures

Climate change adaptation methods for freshwater resources take in water management adaptation measures, including improved water reservoirs in response to expectations of longer periods of drought and encouragement of the use of alternative water sources, for example, reusing wastewater and rainwater harvesting to minimize dependence on freshwater and improve water saving techniques [39]. Other methods also encompass improving irrigation efficiency measures to increase their resilience to climate change and reducing energy needs. Supplementary measures are also viable, i.e. rainwater harvesting system that includes tanks for collecting rainwater from a building's roof, transiting water through filtration pipes and then storing the water and integrating filtered water in feedwater pipes as shown in Fig. 4.14. Moreover, the crop rotations commensurate with the available water quantity and planting of drought-resistant crops to enhance the irrigation infrastructure to reduce water leakage [40].

Additionally, some other solutions include flood defence treatment plants, protecting pipelines from extreme temperatures, using renewable energy in electricity to supply sources to water pumping stations, protecting groundwater resources from salt water intrusion [41], strengthening emergency preparedness and raising public awareness about water saving strategies such as reusing waste water, enhancing efficiency of pipes to prevent water leaks and purchasing water saving products [42].

Fig. 4.15 Greenpix zero-energy media facade. (Source: Arch Daily)

4.1.5 Energy Measures

Underground cabling is used to protect energy infrastructure from floods and hurricane and strengthen energy infrastructure against extreme events with dikes and berms or relocating energy infrastructure in areas outside risk zone [38]. In order to reduce energy demands, some measure could be adapted such as enhancing the buildings' insulation and using energy saving techniques that require minimum energy performance standards for buildings and electricity appliances like lighting fixtures, air conditioners and heaters with labelling and certification programmes for both buildings and key appliances [43]. Figure 4.15 illustrates a zero-energy media facade wall in China, which is also known as the largest coloured display worldwide (as per the publication date of this book). Figure 4.15 also presents the first polycrystalline photovoltaic cell system integrated into a glass curtain wall (35 m high and 60 m wide) and demonstrates innovation and scientific revaluation in building-integrated photovoltaic.

Moreover, measures include the use of renewable sources and diversification of electricity generation sources to reduce the risk during extreme weather events and to upgrade electricity transmission system to maximize the efficiency of electricity distribution [44]. Developing technologies to reduce cooling water in electricity generation plants through the use of dry or hybrid cooling systems, with lower-water requirements, and modifying the number and type of hydroelectricity turbines would assist in adapting to anticipate water flow rates [45].

4.1.6 Health Measures

Measures to be taken in the health sector take account of improving disease, vector surveillance, and monitoring such as mosquito control programme to limit vector disease. In order to adapt to climate change, it is vital to raise public awareness about the emerging threats on health and to enhance efficiency of water supply and sanitation systems, hence, preventing contamination of water supply and monitoring and testing drinking water [46].

Also, extending a state emergency plan, which includes five features: (a) protect the community, especially vulnerable groups by increasing heat alarming systems and awareness raising; (b) reduce impacts of heat stress in urban areas by better urban planning measures to mitigate urban heat islands effect – UHIE; (c) improve the indoor environments through adopting green building codes and practices; (d) enhance emergency management to deal with extreme events; and (e) map the affected zones of extreme events [47]. Other plans could be developed, e.g. air quality management program, developing air quality standards and encouraging eco-friendly transportation, as well as reducing risk of forest fires [48].

4.1.7 Agriculture and Green Space Measures

Specific methods for agriculture and green space measures are developed and increase crop's resistant to heat, drought, disease and pest in order to adapt to water stress and high temperatures. The modernization and diversification of agricultural production and cultivation are considered valuable methods to reduce the risks of flooding and drought yet improve irrigation systems and their efficiency and update servicing and IT systems for prevention of drought damage [49]. Therefore, central governments and local government should develop early warning systems (EWSs) to alarm farmers from extreme events, which could harm their crops. The EWSs also provide climatic forecasts to assist farmers to take decisions related to choosing plant type and planting and harvesting times as well as raising farmers' awareness about opportunities for adaptation measures [50]. Early warning systems have been already installed in some parts in Egypt that has record of sever events. Other methods include improving infrastructure of agriculture, providing large reserves of food raw materials and putting policy measures as well as providing tax incentives and restructuring of the agricultural insurance system to cover rising risks to encourage farmers to assist in adapting to CC [51]. Shade management system (SMS) is to plant crops under canopy of trees to mitigate the effects of extreme temperature and precipitation on crops, e.g. coffee which cannot adapt to high temperatures. Such measure is inexpensive and effective and reduces the ecological and economic vulnerability of many rural farmers in Latin America [52]. Erosion control is the practice of preventing or controlling wind or water erosion in agriculture by physical barriers, such as vegetation or rock, to absorb some of the energy of the wind or water that is causing the erosion [53].

4.1.8 *Forest Measures*

Forest measures are crucial to CC adaptation. Hence, it is vital to incorporate forestry measures, which include: (a) developing sustainable forest management and reforestation (SFMR) programs to change trees species to be drought tolerant types, (b) encouraging better adaptation of species in risk zones, knowing that global temperatures are going to rise by about four degrees over the next 95–100 years as predicted by the Intergovernmental Panel on Climate Change – IPCC, (c) promoting the construction water reservoirs to mitigate droughts or floods, [54] and (d) improving fire management through increasing public awareness, developing contingency plan and training the fire-fighting community [55].

4.1.9 *Biodiversity Measures*

Biodiversity is one of the important measures to CCA. Protecting coastal ecosystem could be achieved by incorporating coastal barriers and natural mechanisms of flood control. Similarly, reducing threats to biodiversity could be through the mitigation of pollution, habitat fragmentation and land and water degradation as well as habitats management [56]. The effective management of natural protected areas (protectorates) and the creation of new protectorates are based on detailed knowledge of the vulnerabilities and sensitivities of species and ecosystems, ecosystem conservation and natural ecosystems restoration [57]. Thus, managing habitats that preserve ecosystem and provide nursery, feeding and breeding grounds for wildlife could be possible through green roofing (Fig. 4.16) [58].

(a) (b)

Fig. 4.16 Green roofs provide nursery, feeding and breeding grounds for birds. (**a**) Green spaces and roof in Incheon, South Korea. (**b**) Green roof in EWA University in Seoul, South Korea. (Source: Lead author)

4.1.10 Economy Measures

Supporting vulnerable groups through adaptation plans should be suitable with the economic and social conditions. Adaption measures that focus on reducing poverty and increasing access to resources will help affected communities or economic sectors [59]. Relocating communities to new lands, when adaptation in their home is not a feasible option, can aid adaptation; however, this choice has negatives impacts on the culture and economics of immigrant and receiving communities [38]. Moreover, measures include expanding insurance to cover CC impacts and transfer risk through climate change workshops and decision-makers' meetings [60]. In Bangladesh, Proshika and Grameen's micro-financing institutions promote loans for the adaptation to climate change by providing safe housing, diversifying incomes in vulnerable sectors and supporting vulnerable communities to help them recover after a disaster [59].

Table 4.2 presents global climate change adaptation measures taken to adapt to climate change risks (CCR) in sectors such as flood-prone and urban areas, infrastructure, energy, water supply, health, agriculture, forestry, ecosystems and economy in Africa, Asia, Europe as well as North America and Latin America.

Table 4.2 and Fig. 4.17 indicate the most of CCA measures are taken in urban areas and infrastructure sector. This shows the seriousness of the damage of CC on cities and indicates that the largest share of the CCA measures and budgets goes to urban sector.

The IPCC Working Group II report pointed out that climate change adaptation (CCA) should be a complementary component of a response strategy to global warming to reduce conflict between measures and increase the ability to integrate adaptation across the full range of development context. Thus, the next chapter investigates CCA policies.

4.2 Conclusions

Chapter 4 supports and cements the notion that climate change (CC) is inevitable and an urgent response and action are crucially needed to reduce the vulnerability of the built environment and related economics to current and future CC risks. Adaptation measures were identified within sectors such as urban areas, infrastructure, coastal areas, freshwater, energy, public health, agriculture, forest, fishing, biodiversity and economy. Results pointed out that most of CCA measures are manifested in urban areas and infrastructure sectors, which indicate seriousness of CC damages on cities globally.

Climate change adaptation (CCA) measures aim at preventing, moderating or assisting in the acclimatise with the consequences of CC risks, for example, early warning systems and vulnerability mapping support in risk assessments and readi-

Table 4.2 (a) Global climate change adaptation measures (1–4)

CCA measures		Continents				
		Africa	Asia	Europe	North America	Latin America
Flood prone	Forecast-warning-response system	✓ [2]	✓ [2]	✓ [61]	✓ [62]	✓ [62]
	Relocating structures away from flood prone	✓ [17]	✓ [63]	✓ [61]	✓ [64]	✓ [65]
	Soft engineering protections such as artificial coral reefs, beach nourishment, green belts	✓ [62]	✓ [66]	✓ [66]	✓ [66]	✓ [66]
	Engineering protections such as sea walls and boulder barriers	✓ [67]	✓ [6]	✓ [6]	✓ [6]	✓ [68]
	Coastal buffer areas	✓ [35]	✓ [62]	✓ [62]	✓ [62]	✓ [62]
	Improve drainage systems	✓ [35]	✓ [62]	✓ [69]	✓ [62]	✓ [70]
Urban Areas	Develop corridor for species migration	✓ [71]	✓ [72]	✓ [73]	✓ [73]	✓ [73]
	Green infrastructure and urban agriculture	✓ [35]	✓ [68]	✓ [35]	✓ [62]	✓ [68]
	Revised building codes and land-use regulations to encourage retrofit and flood-proof structures	✓ [35]	✓ [68]	✓ [35]	✓ [74]	✓ [68]
	Selective relocation	✓ [35]	✓ [68]	✓ [75]	✓ [76]	✓ [68]
	Land use planning and Sustainable urban development	✓ [35]	✓ [59]	✓ [35]	✓ [74]	✓ [68]
	Waste management	✓	✓	✓	✓	✓

Table 4.2 (b) Global climate change adaptation measures (2–4)

CCA measures		Continents				
		Africa	Asia	Europe	North America	Latin America
Urban Areas *(Cont'd)*	Cash transfer programmes	✓ [77]	✓ [77]	✓ [75]	✓ [78]	✓ [77]
	Vulnerability mapping and early warning systems	✓ [68]	✓ [68]	✓ [62]	✓ [76]	✓ [68]
	Building retrofit to increase resilience to CC and enhance water and energy efficiency	✓ [35]	✓ [68]	✓ [35]	✓ [76]	✓ [68]
Infrastructure	Modify road surfaces and railway track to accommodate high temperatures and floods	✓ [68]	✓ [68]	✓ [75]	✓ [74]	✓ [68]
	Protect coastal roads by flood defences and strengthen bridge	✓ [68]	✓ [68]	✓ [75]	✓ [74]	✓ [68]
	Relocate infrastructure outside risk zones	✓ [68]	✓ [62]	✓ [75]	✓ [74]	✓ [68]
	Use eco-friendly transport	✓ [35]	✓ [62]	✓ [75]	✓ [62]	✓ [68]
Energy	Maximize the efficiency of electricity distribution[2]	✓ [68]	✓ [59]	✓ [35]	✓ [74]	✓ [68]
	Strengthening or relocating energy infrastructure against extreme events	✓ [62]	✓ [62]	✓ [62]	✓ [76]	✓ [68]
	Use renewable sources and energy saving techniques	✓ [68]	✓ [59]	✓ [62]	✓ [76]	✓ [68]
Water supply	water demand management	✓ [35]	✓ [62]	✓ [62]	✓ [38]	✓ [62]

✓ Mark means that the measure is adopted in regional policies

[ ] The red number between two red bracts is reference

Table 4.2 (c) Global climate change adaptation measures (3–4)

CCA measures		Continents				
		Africa	Asia	Europe	North America	Latin America
Water supply *(Cont'd)*	Reuse wastewater and rainwater harvesting	✓ [35]	✓ [62]	✓ [62]	✓ [79]	✓ [74]
Water supply *(Cont'd)*	Improve irrigation efficiency and use drought-resistant crop	✓ [62]	✓ [62]	✓ [62]	✓ [62]	✓ [62]
Water supply *(Cont'd)*	Enhance water infrastructure and water reservoirs	✓ [35]	✓ [62]	✓ [35]	✓ [62]	✓ [62]
Water supply *(Cont'd)*	Protect groundwater from saltwater intrusion	✓ [35]	✓ [80]	✓ [81]	✓ [82]	✓ [83]
Water supply *(Cont'd)*	Raise public awareness about water saving strategies	✓ [84]	✓ [84]	✓ [35]	✓ [85]	✓ [86]
Health	Improve disease and vector surveillance and monitoring	✓ [35]	✓ [59]	✓ [35]	✓ [87]	✓ [62]
Health	Improve outdoor climate and public health services	✓ [35]	✓ [59]	✓ [35]	✓ [76]	✓ [62]
Health	Develop early warning systems	✓ [59]	✓ [59]	✓ [62]	✓ [74]	✓ [62]
Health	Raise public awareness	✓ [59]	✓ [59]	✓ [75]	✓ [74]	✓ [59]
Agriculture	Educate farmers use of new adaptation techniques	✓ [59]	✓ [59]	✓ [38]	✓ [74]	✓ [62]
Agriculture	Develop of early warning systems	✓ [59]	✓ [88]	✓ [62]	✓ [89]	✓ [59]
Agriculture	Improve irrigation systems	✓ [35]	✓ [68]	✓ [35]	✓ [74]	✓ [59]
	climatic forecasts in agricultural planning	✓ [59]	✓ [88]	✓ [90]	✓ [89]	✓ [62]

✓ Mark means that the measure is adopted in regional policies

[ ] The red number between two red bracts is reference

Table 4.2 (d) Global climate change adaptation measures (4–4)

CCA measures		Continents				
		Africa	Asia	Europe	North America	Latin America
Agriculture *(Cont'd)*	Develop crops resistant to heat, drought and disease	✓ [59]	✓ [68]	✓ [62]	✓ [74]	✓ [62]
	Erosion control	✓ [59]	✓ [59]	✓ [75]	✓ [91]	✓ [59]
	Changes in planting and harvesting times	✓ [35]	✓ [59]	✓ [35]	✓ [91]	✓ [59]
	Crop rotation and livelihood diversification	✓ [62]	✓ [62]	✓ [62]	✓ [92]	✓ [92]
	Modernize agricultural production and cultivation methods	✓ [59]	✓ [88]	✓ [62]	✓ [91]	✓ [59]
Forestry	Improved wild management	✓ [93]	✓ [93]	✓ [62]	✓ [74]	✓ [93]
	Reforestation	✓ [93]	✓ [93]	✓ [62]	✓ [74]	✓ [93]
Ecosystems	Designing marine protected areas	✓ [62]	✓ [88]	✓ [94]	✓ [91]	✓ [95]
	Reducing pollution, land and water degradation and habitat fragmentation	✓ [68]	✓ [59]	✓ [62]	✓ [76]	✓ [59]
	Managing habitats that preserve ecosystem	✓ [35]	✓ [88]	✓ [35]	✓ [35]	✓ [68]
Economy	Reduce potential economic effects of climate on the local economy	✓ [68]	✓ [68]	✓ [62]	✓ [74]	✓ [68]
	Support green economy	✓ [35]	✓ [88]	✓ [62]	✓ [91]	✓ [96]
	Government insurance coverage CC risks	✓ [62]	✓ [88]	✓ [62]	✓ [74]	✓ [59]
Total of CCA Measures		49	49	49	49	49

✓ Mark means that the measure is adopted in regional policies

[ ] The red number between two red bracts is reference

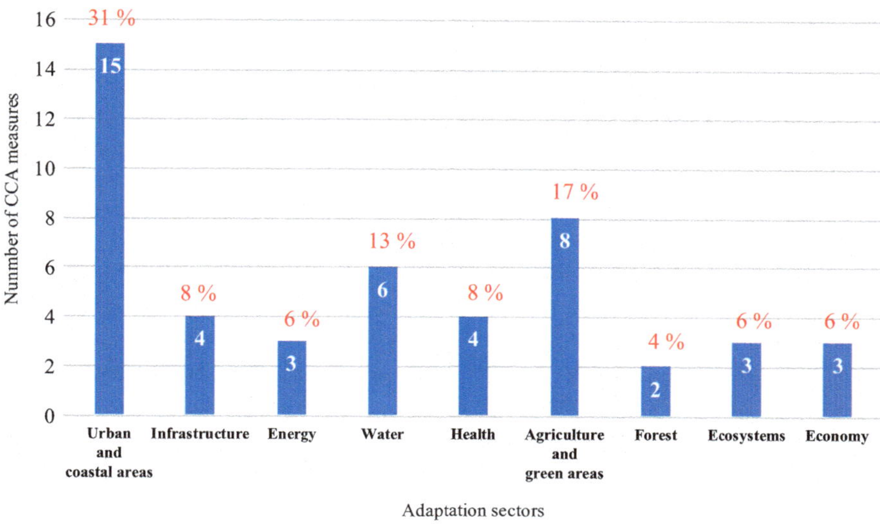

Fig. 4.17 Number of climate change adaptation (CCA) measures by sector. (Source: Developed by authors)

ness for and resilience of the built environment and infrastructure to cope with disasters.

The chapter presented the CCA measures in coastal areas that are taken through protection by engineering or ecological measures, accommodation by flood-proofing measures or retreat by moving property away of flood prone. Also, climate adaptation measures in urban areas including several examples were listed:

(a) Green infrastructure works as sink for storm water and also reduces heat island effect.
(b) Waste management through waste minimization, reuse, recycling, converting waste to energy or disposal of waste and relocation of waste routes away from flood plains.
(c) Development of building regulations, e.g. flood-proof structures and green building codes, and encouragement of thermal retrofitting and water and energy efficiency in buildings to integrate CC impacts into design.
(d) Imposing land use restriction in flood-prone areas such as limited low-density land activities and reducing dependencies on transportation and developing corridors to connect wildlife habitats.

In addition, this chapter depicted ways to adapt infrastructure, including using heat resistance materials, raising roads, strengthening bridges and upgrading track-side drains or using porous pavement to adapt to floods. Adaptation for fresh water could be achieved by reusing wastewater, rainwater harvesting, improving water and irrigation efficiency. Adaptation measures for energy sector is enhanced by strengthening energy infrastructure and depending on renewable energy, energy-saving techniques, hybrid cooling systems and modern hydroelectricity turbines to adapt to droughts. Adaptation measures in the health sector were presented and discussed. These measures were achieved through three steps: controlling vector disease, improving the indoor climate through green building codes and improving air quality by encouraging eco-friendly transportation. Moreover, adapting agriculture is another CCA measure, which was achieved by shade management systems (SMS) that mitigate the effects of extreme temperatures and using physical barrier as vegetation or rock to control wind erosion. Furthermore, adaptation of ecosystem is pursued by reducing pollution, habitat fragmentation, land degradation and habitats conservation and restoration, as well as saving the economy through supporting vulnerable groups by reducing poverty, increasing access to resources and expanding insurance to cover CC impacts.

Finally, urban and coastal areas are the highest sectors in terms of number of CCA measures, 31 per cent followed by agriculture and water 17 per cent and 13 per cent, respectively, whereas other measures (infrastructure and health) 8 per cent; energy, ecosystem and economy each 6 per cent; and forest as low as 4 per cent.

It is vital to note that the management and assessment of CC measures in a large scale, such as a city or a country, need to integrate these measures into policies related to disaster risk reduction, coastal management, urban development and spatial planning, agriculture health services as well as ecosystems, energy and water management. The outcome of managing CC adaptation through policies is better than measures. Additionally, it is significant to lessen conflict between measures and increase the ability of integrating adaptation across the full range of development context. Therefore, the next chapter defines, presents and discusses *climate change adaptation policies* and their framework structure.

References

1. William R. Travis, "Elements of a Severe Climate Change Early Warning System" [Online]. Available at: http://www.heardisland.org/HD_documents/HE_Library/Climate/Travis_warning_system.pdf. (Accessed: March 10, 2015).
2. "Strengthening Climate Information and Early Warning Systems for Climate Resilient Development" United Nations Development Programme, [Online]. Available at: http://www.undp-alm.org/ews-and-climate-resilient-development. (Accessed: July 28, 2015).
3. Jon Barnett Michael Webber, "Accommodating Migration to Promote Adaptation to Climate Change", The World Bank Development Economics Office, 2010.
4. Laura Tam, "strategies for Managing Sea Level Rise" *THE URBANIST,* no. 487, 2009.
5. "Survey of Climate Change Adaptation Measures in Maldives", Ministry of Housing and Environment, 2011.
6. Matthew M. Linham and J. Nicholls, "Technologies for Climate Change Adaptation" United Nations Environment Programme, November 2010.
7. Yumi Lee, "Coastal Planning Strategies for Adaptation to Sea Level Rise: A Case Study of Mokpo, Korea", *Journal of Building Construction and Planning Research,* pp.74–81, 2014.

8. Keith H. Hirokawa and Jonathan D. Rosenbloom, "Land Use Planning in a Climate Change Context", Edward Elgar, 2013.
9. Jessica Grannis, "Adaptation Tool Kit: Sea-Level Rise and Coastal Land Use", Georgetown Climate Center, 2011.
10. Norval Collins, "A Guide for Incorporating Adaptation to Climate Change into Land Use Planing" MCIP CEF Consultants, November 2005.
11. "Awareness campaigns for behavioural change" European Climate Adaptation Platform, [Online]. Available at: http://climate-adapt.eea.europa.eu/viewmeasure?ace_measure_id=656. (Accessed: July 18, 2015).
12. Erin L. Bohensky, Alex Smajgl and Tom Brewer, "Patterns in household-level engagement with climate change in Indonesia", *Nature Climate Change,* pp.348–351, 2013.
13. UNEP Regional Seas Reports and Studies, "Pacific Island Mangroves in a Changing Climate and Rising Sea", United Nations Environment Programme, 2006.
14. Martin Ely and Sheryn Pitman, "Green Infrastructure Life support for human habitats The compelling evidence for incorporating nature into urban environments", Green Infrastructure Project, Botanic Gardens of South Australia, 2014.
15. Intergovernmental Panel on Climate Change, "Climate change and water Intergovernmental Panel on Climate Change climate change and water IPCC Technical Paper VI" Intergovernmental Panel on Climate Change, June 2008.
16. Mark Snow and Deo Prasad, "Climate Change Adaptation for Building Designers: An Introduction", Australian Institute of Architects, 2011.
17. "Adapting to climate change in Africa", *Joto Afrika,* no. 3, April 2010.
18. Anton Cartwright, "Global Climate Change and Adaptation – A Sea-Level Rise Risk Assessment", The City of Cape Town Environmental Resource Management Department, July 2008.
19. "Solid Waste Management Addressing Climate Change Impacts on Infrastructure: Preparing for Change", U.S. Agency for International Development (USAID), 2012.
20. The World Bank Group, "Solid Waste, guide to Climate Change and Adaptation in cities" The World Bank Group, [Online]. Available at: www-esd.worldbank.org/citiesccadaptation/solid-waste.html. (Accessed: July 28, 2015).
21. "Sustainable Urban Development Planner for Climate Change Adaptation" European Union, [Online]. Available at: https://climate-adapt.eea.europa.eu/metadata/projects/sustainable-urban-development-planner-for-climate-change-adaptation. (Accessed: July 28, 2015).
22. The Energy and Resources Institute (TERI), "Adaptation to Climate Change in the Context of Sustainable Development Background Paper prepared under contract By, United Nations Department of Economic and Social Affairs Division for Sustainable Development Cli" in *Climate Change and Sustainable Development*, New Delhi, 7–8 April 2006.
23. "Developing Early Warning Systems: A Checklist" in *EWC III Third International Conference on Early Warning From Concept to Action*, Bonn, Germany, 27–29 March 2006.
24. Rasmus Heltberg and Misha Bonch-Osmolovskiy, "Mapping vulnerability to climate change", The World Bank, January 2011.
25. "Climate Change Monitoring System" climate tech wiki, [Online]. Available at: http://www.climatetechwiki.org/content/climate-change-monitoring-system#Advantages of the technology. (Accessed: July 16, 2015).
26. Chido R Mutambara, "An Evaluation of Cash transfer Programmes in Zimbabwe as a Strategy to Reduce Poverty" University of Zimbabwe, October 2011.
27. Mark Davies, Katy Oswald and Tom Mitchell, "Climate Change Adaptation, Disaster Risk Reduction and Social Protection" The Organisation for Economic Co-operation and Development (OECD), 2009.
28. NHS Sustainable development Unit, "Adaptation to Climate Change for Health and Social Care organisations" NHS Sustainable development Unit, 2012.
29. Chris Reardon and Dick Clarke, "Passive cooling" [Online]. Available at: www.yourhome.gov.au/sites/prod.yourhome.gov.au/files/pdf/YOURHOME-2-PassiveDesign-5-PassiveCooling-(4Dec13).pdf. (Accessed: August 18, 2015).

30. "Flood-proofing" climate tech wik, [Online]. Available at: www.climatetechwiki.org/content/flood-proofing. (Accessed: August 19, 2015).
31. Jon Braman, Steven Kolberg and Jeff Perlman, "Energy and Water Savings in Multifamily Retrofits", Bright Power, Inc., 2014.
32. The Soil Resources, Management and Conservation Service, "Guidelines for land-use planning" Food and Agriculture Organization of the United Nations, Rome, 1993.
33. Gregory R.A. Richardson and José Otero, "Land use planning tools for local adaptation to climate change" Her Majesty the Queen in Right of Canada, 2012.
34. Bhishna Bajracharya, "Climate change adaptation through land use planning and disaster management: Local government perspectives from Queensland" in *17th Pacific Rim Real Estate Society Conference Climate change and property: Its impact now and later,* January 2011.
35. Aromar Revi et al, "Climate Change 2014: Impacts, Adaptation, and Vulnerability" in *Chapter 8. Urban Areas*, Cambridge, UK, Cambridge University Press, March 31, 2014, pp.535–612.
36. Øystein Hov, Ulrich Cubasch and others, "Extreme Weather Events in Europe: preparing for climate change adaptation", Norwegian Meteorological Institute, October 2013.
37. "Infrastructure, Engineering and Climate Change Adaptation – ensuring services in an uncertain future", The Royal Academy of Engineering on behalf of Engineering the Future, 2011.
38. "Climate Change 2007: Synthesis Report" Intergovernmental Panel on Climate Change, 2007.
39. Economic Commission for Europe, "Guidance on Water and Adaptation to Climate Change", United Nations Publications, 2009.
40. Copa Europe farms, "Water and agriculture under a changing climate" Copa Europe farms, [Online]. Available at: www.copa-cogeca.be/img/user/file/Climate/5660%20version%20E.pdf. (Accessed: July 17, 2015).
41. Kommunalkredit Public Consulting GmbH, "Improving the Environmental Quality of the Black Sea through Better Waste Water Treatment & Climate Change Adaptation of the Water Sector in Moldova", European Commission, 2013.
42. US Environmental Protection Agency, "Fact Sheet: 21 Water Conservation Measures for Everybody" US Environmental Protection Agency, [Online]. Available at: http://nepis.epa.gov/Exe/ZyPDF.cgi/10003HYV.PDF?Dockey=10003HYV.PDF. (Accessed: January 10, 2016).
43. "Climate Risk and Adaptation in the Electric Power Sector" Asian Development Bank, 2012.
44. "Adaptation Examples: Energy" US Environmental Protection Agency, [Online]. Available at: www.epa.gov/climatechange/impacts-adaptation/energy-adaptation.html. (Accessed: August 18, 2015).
45. "Water in a Low-Carbon Economy: Resource Scarcity, Climate Change and Business in a Finite World, POLICY BRIEF", Stockholm Environment Institute, 2012.
46. "Adaptation in Action: Grantee Success Stories from CDC's Climate and Health Program" American Public Health Association, March 2015.
47. Sara P. Hoverter, "Adapting to Urban Heat: A Tool Kit for Local Governments" Georgetown Climate Center, 2012.
48. "Health impacts of climate change: Adaptation strategies for western Australia" Environmental Health Directorate, Department of Health, 2008.
49. Rebecca Clements, "Technologies for Climate Change Adaptation – Agriculture Sector" UNEP Risø Centre on Energy, Climate and Sustainable Development, 2011.
50. P.A. Jaranilla-Sanchez et al., "A Primer on Climate Change Adaptation in the Philippines", World Agroforestry Centre, 2007.
51. Ada Ignaciuk and Daniel Mason-D'Croz, "Modelling Adaptation to Climate Change in Agriculture", OECD Publishing, 2014.
52. Laurence Jassogne, Peter Läderach and Piet van Asten, "The Impact of Climate Change on Coffee in Uganda: Lessons from a Case Study in the Rwenzori Mountains", Oxfam Research Reports, 2013.
53. R.P. Singh, "Environmental Studies", Concept Publishing Company, 2012.

54. Ad Hoc Working Group III on Climate Change and Forestry, "Climate Change and Forestry Report", The Standing Forestry Committee, 2010.
55. Samuel Mulenga, "Opportunities for Adaptation and Mitigation through Africa Bio-Carbon Initiative", Center for International Forest Research, Zambia, 2010.
56. Sandra Naumann and Gerardo Anzaldua, "Assessment of the potential of ecosystem-based approaches to climate change adaptation and mitigation in Europe" European Commission, November 2011.
57. The Canadian Parks Council Climate Change Working Group, "Canadian Parks and Protected Areas: Helping Canada weather climate change", Parks Canada Agency on behalf of the Canadian Parks Council, 2013.
58. "Natural Resource Program", New York State Department of State, April 2010.
59. "Impacts, Vulnerabilities and Adaptation in Developing Countries", United Nations Framework Convention on Climate Change (UNFCCC), 2007.
60. Graham Sem, "Vulnerability and Adaptation to Climate Change in Small Island Developing States" UNFCCC, [Online]. Available at: unfccc.int/files/adaptation/adverse_effects_and_response_measures_art_48/application/pdf/200702_sids_adaptation_bg.pdf. (Accessed: August 19, 2015).
61. "Floods: Climate Change and Adaptation Strategies for Human Health" WHO, London, UK, 2002.
62. The Intergovernmental Panel on Climate Change (IPCC), "Climate Change 2014: Impacts, Adaptation and Vulnerability", Cambridge University Press, Cambridge, UK, 2014.
63. Jane M. Chun, "Planned Relocations in the Mekong Delta: A Successful Model for Climate Change Adaptation", Brookings Institution, 2015.
64. "Adaptation Examples: Coastal Areas" EPA, [Online]. Available at: www.epa.gov/climatechange/impacts-adaptation/coasts-adaptation.html. (Accessed: November 23, 2015).
65. "Climate Change Adaptation Planning in Latin American and Caribbean Cities" ICF GHK in association with King's College London and Grupo Laera Job Number, October 2012.
66. Narayan S., et al., "The Effectiveness, Costs and Coastal Protection Benefits of Natural and Nature-Based Defences" *PLoS ONE,* 2016.
67. Ecologic Institute and AEA group, "Adaptation to Climate Change Policy instruments for adaptation to climate change in big European cities and metropolitan areas" European Union, 2011.
68. Hunt and Watkiss, "Sea level rise and coastal flood risk, health and water resources" THe World Bank, 2011.
69. Joanne Langis Groupe, "Adaptation measures for floods, storm surges, and sea level rise" Université de Moncton, 2013.
70. "Climate Change Adaptation Planning in Latin American and Caribbean Cities" The World Bank Group, 2015.
71. Caroline Petersen, "South Africa: Ecosystem-Based Planning for Climate Change", South African National Biodiversity Institute, Washington, DC.
72. Case Studies Illustrating the Socio-Economic Benefits of Ecological Networks, Montreal: Secretariat of the Convention on Biological Diversity, 2010.
73. Sarah-Taïssir Bencharif, "Climate Change And Connectivity: Are Corridors the Solution?" Queen's University Kingston, Ontario, Canada, 2010.
74. Ligia Castro et al, "Ch:12 Latin America and the Caribbean" in *Global Environment Outlook (GEO-5),* UNEP, 2012.
75. "Urban adaptation to climate change in Europe" European Environment Agency, Copenhagen, 2012.
76. Rachel M. Gregg et al., "The State of Marine and Coastal Adaptation in North America: A Synthesis of Emerging Ideas Final Report" EcoAdapt, January 2011.
77. laine Fultz and John Francis, "Cash transfer programmes, poverty reduction and empowerment of women: A comparative analysis Experiences from Brazil, Chile, India, Mexico and South Africa" International Labour Organization, 2013.

78. "Details on Transfer Payment Program" Government of Canada, [Online]. Available at: www.ec.gc.ca/default.asp?lang=En&n=5F60D654-1&offset=4&toc=show. (Accessed: December 14, 2015).
79. Paul Moss et al., "eAdapting to Climate Chang in Minnesota 2013" The Interagency Climate Adaptation Team, November 2013.
80. UNESCO, "Groundwater Management Coastal areas" Asian Development Bank, [Online]. Available at: http://adb-knowledge-partnership.unesco-ihe.org/comparative-research-groundwater-management-0. (Accessed: November 21, 2015).
81. "Adaptation of groundwater management (2015)" climate-adapt, [Online]. Available at: http://climate-adapt.eea.europa.eu/viewmeasure?ace_measure_id=636. (Accessed: November 22, 2015).
82. Salt Water Intrusion Barrier Wells, The Class V Underground Injection Control Study, United States Office of Ground Water, EPA, 1999.
83. "Regional Diagnostic Report Latin America and the Caribbean", in *Groundwater Governance - A Global Framework for Action*, Montevideo, Uruguay, 18–20 April 2012.
84. "Social Aspects of Sustainable Development in South Africa" The Government of South Africa to the fifth session of the United Nations Commission on Sustainable Development, 1997.
85. "How Efficiency Programs Help Water Utilities Save Water and Avoid Costs", United States Environmental Protection Agency Office of Water, 2002.
86. "Implementation of the water policy and strategy of the United Nations Environment Programme" Governing Council of the United Nations Environment Programme, 2009.
87. "Public Health Agency of Canada 2014–2015, Departmental Performance Report", The Honourable Jane Philpott, 2015.
88. "The Climate Change Commission" Office of the President of Philippines, [Online]. Available at: http://climate.gov.ph/index.php/the-ccc. (Accessed: March 12, 2016).
89. "Case Studies" The U.S. Climate Resilience Toolkit, [Online]. Available at: https://toolkit.climate.gov/taking-action?f[0]=field_climate_stressor%3A103&f[1]=field_parent_topic%3A115. (Accessed: November 14, 2015).
90. Marta Bruno Soares and Suraje Dessai, "Exploring the use of seasonal climate forecasts in Europe through expert elicitation", *Elsevier,* vol. 10, pp. 8–16, 2015.
91. "Climate Change Adaptation Plan", US Environmental Protection Agency, 2014.
92. James J. McCarthy et al., "Climate change 2001: Impacts, Adaptation and Vulnerability" Cambridge University Press, Cambridge, UK, 2001.
93. "The UN-REDD Programme" UN-REDD, [Online]. Available at: www.un-redd.org/aboutredd. (Accessed: September 10, 2015).
94. "Marine protected areas in Europe's seas", European Environment Agency, Copenhagen, 2015.
95. Guerrero, E. et al., Protected Areas and Development in Latin America - From Santa Marta 1997 to Bariloche 2007 and Perspectives for a New Decade, Bogotá: IUCN Colombian Committee, Parques Nacionales Naturales Colombia and Fundación Natura, 2009.
96. "XXXVIII Regular Meeting of the Latin American Council" in *The vision of the green economy in the Latin America and the Caribbean*, Caracas, Venezuela, 17–19 October 2012.

Chapter 5
Global Climate Change Adaptation Policies

Abstract Adaptation policies are seamlessly crucial in the efforts to lessen climate change risks. This chapter consists of three parts; the first part investigates climate change adaptation (CCA) policies that were adopted in Africa, Asia, Europe, North America and Latin America. These CCA policies are related to urban areas, infrastructure, water resources and energy as well as green infrastructure, public health and tourism sectors. These adaption policies have a significant role in buildings' retrofitting and cities' upgrading. The second part identifies the mechanism of forming and prioritising CCA policies to be integrated into the climate action plans. The third part determines the challenges and barriers appeared during adaptation process; thus, it suggests the method to overcome these obstacles for better adaptation practices. The importance of this chapter stems from discussing CCA policies that are responsible for managing CC measures in a large scale, whether on a city level or a country level to avoid conflict with other measures and developing plans and facilities that can be integrated into existing policies. Nonetheless, by adopting policies only, this doesn't seem useful; thus, discussing tools such as adaptation decision framework is vital to shove adaptation from theory phase to practice and yet learn how to overcome the obstacles it faces for successful climate change adaptation.

5.1 Global Climate Change Adaptation Policies

Climate change adaptation (CCA) policy is defined as the sum of processes pursued by the government or administration, which aims at influencing adaptation to climate change (CC) by avoiding or reducing the adverse consequences of risks and taking advantage of new opportunities resulting from CC. Policies include climate action plans, national adaptation, support programmes and pilot activities and financing instruments as well as laws and regulations, sectors' strategies, plans and measures and formal and informal working groups in addition to other institutional or governance arrangements [1]. In 1995, the IPCC published the CC impacts and adaptations' guideline, which represents the first generation of adaptation decision framework to assess CC impacts and select appropriate adaptations decision. In a

M. M. Aboulnaga et al., *Urban Climate Change Adaptation in Developing Countries*, https://doi.org/10.1007/978-3-030-05405-2_5

global context, many important steps have been also taken in the development of the adaptation regime under the UNFCCC in 2001 at the seventh session of the Conference of the Parties (COP 7) in Marrakesh. In addition, the National Adaptation Programmes of Action (NAPA) was established to identify and report the adaptation actions needed for the least developed countries, thus it highlights the importance of adaptation policies that play a major role in buildings' improvements in energy performance and efficiency as well as in cities' upgrading through retrofitting and resilience measures.

The next section discusses climate change adaptation policies adopted in Africa, Asia, Europe, North America and Latin America for coping with climate change risks in urban areas, infrastructure, water resources, energy, green infrastructure, public health and tourism sector.

5.1.1 Urban Areas Policies: Coastal Zones

The climate change adaptation strategy used for coastal areas in Europe is presented in Table 5.1. It is divided into three parts: (a) grey such as passive cooling, (b) green such as urban areas and (c) soft adaptation measures including plans and actions that are related to reducing the climate change impacts. Adaptation responsibilities are distributed between national government and local and regional governments. The national government responsibility is summarized in assessing the vulnerabilities of the national development plans, the impacts of climate change and monitoring adaptation process in order to develop early warning systems and methods to raise community awareness about risks, law and regulations and strategic frameworks in response to climate change. Local and regional governments' responsibility is summarized in integrating climate change within the regional strategies and research programmes and financing small and medium regional adaptation projects [2].

In order to create more resilient plans to climate change (CC) impacts, governments should integrate adaptation objectives, sustainable development objectives and climate change considerations into planning. Also, governments should investigate the sensitivity of development plans with respect to being enhanced with suitable adaptation measures to be more resilient to CC impacts. Moreover, governments should design adaptation policy guidelines in order to evaluate the ability of urban areas' policies to integrate with changing adaptation projects. Additionally, governments are required to set up a CC information management, gather CC-related data and analyse specific risks as well as support CC researches in order to enable smooth information sharing and support responses of decision-makers and stakeholder [3].

Engaging public, business, and other stakeholders' groups in the development of adaptation strategies is essential. This could be done by raising awareness among communities regarding CC impacts that their city faces and informing residents about adaptation plans, policies and actions adopted by the city to meet CC impacts and to ensure their response to CC risks. Stakeholders' involvement should assist in the decision-making process by expressing their ideas, inputs and suggesting actions that can be taken by individuals and groups [4].

Table 5.1 Adaptation measures to flooding in cities (EEA, 2012)

Grey measures	Green measures	Soft measures
Make new buildings and infrastructure flood proof by appropriate design and material use Maintenance/upgrade of drainage system Temporary water storage in basins or fascines Separate treatment of rain water, disconnected from sewage, improved ground drainage Innovative design of buildings and areas such as elevated entrances, building on poles, floating houses, temporary water storage, green roofs Dams, flood defences	Avoid/remove impervious surfaces wherever possible Maintain and further increase green infrastructure in cities, parks and gardens, wetlands, water bodies and also green roofs Maintain and manage green areas outside and inside the cities for flood retention including the use of appropriate agricultural and forest practices 'Re-naturalisation' of rivers and wetlands	Mapping of flood risks taking into account climate change scenarios and information distribution Forecasting and early warning systems Awareness raising, knowledge and capacity building for all groups to cope with floods and flood risks Strategic planning in river basins – ban building in flood-prone areas; protect flood retention and other green areas Flood risk management plans Rain water management Guidance for behaviour changes such as not storing valuables in basements Adapting building and planning codes to include flood resistance Taxes or incentives, such as concerning the amount of sealed area per property, amount of waste water used (including rain water) Insurance of damages

Integrated storm water management plans should be developed to adapt to flooding by decreasing the leakage of storm water through applying storm water management techniques, increasing storm sewer capacity through separating storm sewer from sanitary infrastructure and enhancing green space to improve rainwater retention [5]. Also, land use regulation should take into account allocating proper land use in areas that are less vulnerable to CC risks. Moreover, transportation should be coordinated to save lives and economy in cases of emergency as well as promoting green infrastructure and watershed management to reduce storm water runoff. Another solution is rehabilitating coastal mangroves to provide habitat and flood protection and monitor groundwater salinity in coastal zones [6].

Also, it is important to mitigate the urban heat island effect (UHIE) thus improving public health and minimizing heat stroke and mortality during heat waves. This could be achieved through planning and expanding the hot weather preparedness work programme, which includes UHIE and vulnerable population mapping that provide cooling systems in civic facilities and increase the number of green infrastructure [7]. In addition, other methods can be implemented such as strengthening building codes' regulations by including insulations and shading techniques to increase energy saving, minimize heat gain and improve the durability of structure to adapt with floods and cyclones [8].

Similarly, other measures include: (a) the development of policies for backup power in major and large public buildings; (b) assessment of shortfalls in energy network during extreme weather events [9]; (c) providing temporary shelter to vulnerable people in emergency, such as extreme weather events, through the use of public buildings, i.e. schools or community centres, as part of the disaster response plan; and (d) extending insurance coverage to include climate change and natural disaster [10]. Lastly, decreasing damage from forest fires can be attained through training fire-fighters to deal with forest fires and providing them with special equipment. Also, improving fire-fighters' access to risk areas during emergency by enhancing roads' network that permeate into forest areas, and installing watchtowers and camera systems, as well as early warning systems to assist in detecting fires at an early stage can mitigate the risk in such areas and enhance resilience. In addition, delaying infrastructure adaptation actions will increase costs at a later time because the replacement cost of infrastructure (buildings, roads, railways, energy grids and sewage systems) needs a large financial budget. Thus, adapting the city with future CC impacts and preventing building on prone areas will result in lower costs and increased effectiveness [2].

5.1.2 Infrastructure Policies

Cities' authorities and local governments should work on assessing and evaluating the vulnerability of existing infrastructure to CC future scenarios, then identifying the vulnerability factors that increase CC impacts and determining the challenges and barriers that face the adaptation process, hence making these cities resilient.

Also, estimating the costs and benefits of adaptation actions compared to the costs of no action is vital [3]. In addition, cities' authorities and local governments should act on incorporating adaptation into national transport policies and in the planning, design, operations and maintenance of city's infrastructure and set recommendations to identify risk areas such as flood-prone areas as well as making the transport and mobility clean, smart and resilient. Cities should also estimate the severity of CC impact to address vulnerabilities and adaptive measures, including enhancing drainage, structures and porous pavement to adapt to floods and treating the soil in order to adjust with landslides and adding some additions to improve the asphalt mixture's resistance to heat and water [11]. Moreover, cities should improve urban street design strategies by encouraging eco-friendly transport, walking, and cycling as well as providing public open spaces and natural green corridors to provide cooling. Furthermore, cities' authorities can develop thermally comfortable public transportation by installing air cooling systems to stations, repainting to white to reflect heat and using high reflectivity glass and thermostatically controlled heating systems to save energy [12]. Figure 5.1 illustrates an example of a street model in Chicago that addresses a variety of strategies bundles such as light, porous pavement, landscape and recycling as well as eco-transport service transit, cars, cyclists and pedestrians. A vivid example is manifested in the city of Barcelona, Spain, where the city was enhanced by urban streets' design strategies that are encouraging eco-friendly transport, walking and cycling by providing public open spaces and natural green corridors to provide cooling (Fig. 5.2).

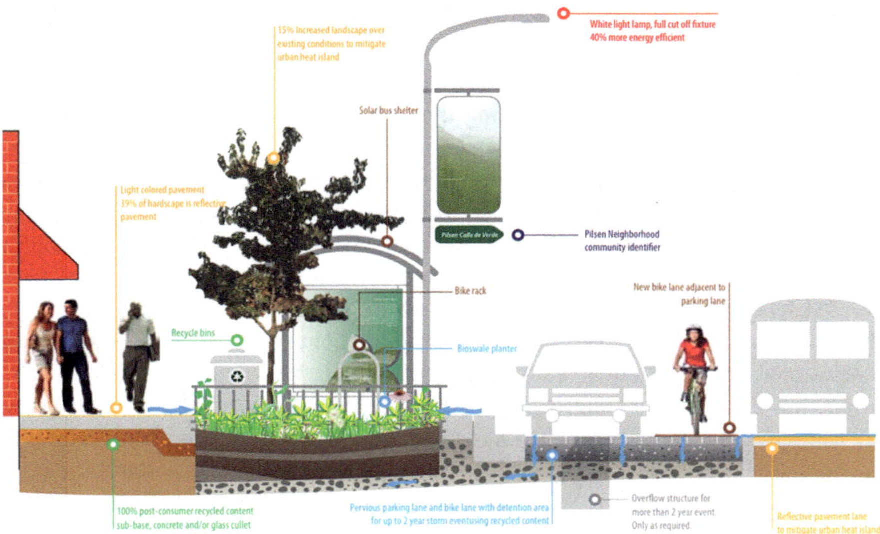

Fig. 5.1 West Cemark Street plan, Chicago. (Source: U.S. Geological Survey)

(a) (b) (c)

Fig. 5.2 Diagonal street in Barcelona, Spain

5.1.3 Water Resource Policies

Water resource management policies are used to guarantee the resilience of CC adaption and preserve water resources for future generations. Water management policies are divided into soft adaptation and hard adaptation strategies. Soft adaptation strategies focus on regulations and operations to reduce barriers and create opportunities, while hard strategies focus on engineering and technology such as building water reservoirs and flood defences and improving irrigation systems to preserve human and ecosystem health [13]. Policies aim to reducing current vulnerability to droughts and floods, closing the demand-supply gap and restoring and maintaining water resources [14]. Policies include integrating adaptation to CC into water resource management by incorporating adaptations measures in development plans and sectorial strategy plans such as agriculture, energy and urban areas and monitoring relevant programmes [15]. Climate change adaptation policies include the development of emergency response plans to deal with the impacts of sea level rise (SLR) on land use and water sectors and the engagement of stakeholders into the adaptation process by raising awareness about risks and measures [16]. Operational strategies aim at reducing the gap between water supply and demand by protecting groundwater, improving water use efficiency in agriculture and urban areas, increasing storage capacity and expanding the use of economic incentives to encourage sustainable water use [17]. Infrastructure strategies aim at increasing infrastructure responses to SLR and storms and floods by considering extreme weather events in evaluating water infrastructure safety, retrofitting water and sewage in order to reduce water runoffs and flooding to recycle wastewater and water storage [3].

5.1.4 Energy Policies

Enhancing the energy sector could be achieved by the diversification of energy sources and increasing use of off-grid alternative energy resources such as solar, wind and geothermal energy. Hence, it is important to implement the following strategies: (a) encouraging the development of renewable resources, (b) developing energy efficient technologies in household, public and industrial sectors, and (c) encouraging stakeholders to install small electricity networks grid by giving them financial incentives for linking their networks to the public electricity grid [18]. In terms of climate change adaptation measures in the energy sector, the following seven measures should be included in the adaptation plan: (a) developing early warning and response systems for energy to react quickly to post-disaster recovery needs, (b) encouraging energy conservation, (c) mapping vulnerable areas to CC impacts and assessing their adaptation ability [19], (d) assessing environmental impacts in the setting, and (e) relicensing of new energy infrastructure along with upgrading power plants to decrease water consumption, increasing resilience to high temperatures, as well as (f) developing hydroelectricity turbines to adapt to decreased water flow, and (g) using adaptation engineering measures to protect energy infrastructure from extreme weather events [20]. Figure 5.3 shows an example of adaptation actions such as increasing the water reservoir capacity to adapt to drought and strengthening railway, power pylons and bridges to be anti-flood.

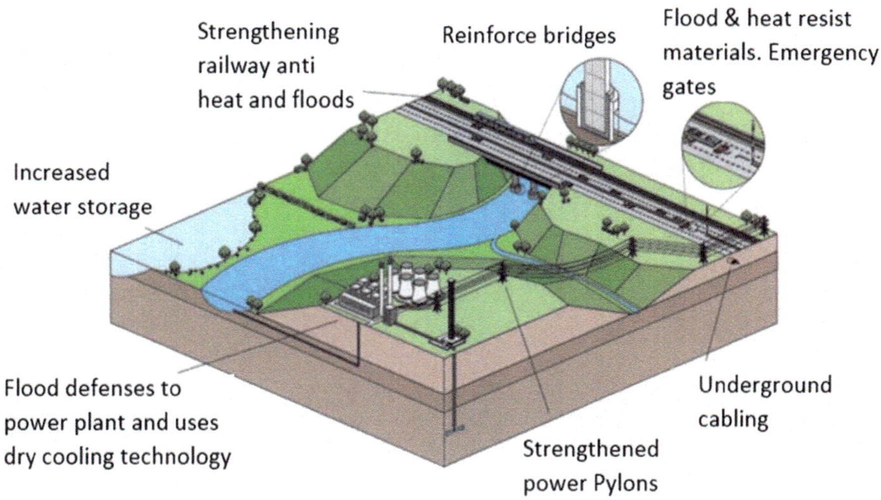

Fig. 5.3 Adaptation actions in infrastructure sector. (Source: The Secretary of State for Environment, UK)

5.1.5 *Agriculture and Green Infrastructure Policies*

It is essential to integrate climate change adaptation into existing food security's strategy, national plans and research institutions through research and development (R&D) [17]. Policies are creative work based on systematic basis and aim at increasing agricultural research about crops, forestry and fisheries taking into consideration the socio-economic aspects. Agriculture research topics including farm crop storage, processing of agricultural products and food processing will create new applications in the agricultural sector that will identify and monitor CC impacts [21].

Sustainable management of water utilization in agriculture could be achieved by developing irrigation and water management systems to reduce water losses and enhance the resistance of crops to heat, drought and disease [22]. Other methods include the protecting and raising the efficiency of agricultural soil against droughts, desertification and floods by applying sustainable management, developing cultivation methods, raising awareness of farmers and issuing land tenure regulations to increase resilience to CC [23]. Also, it is important to limit problems such as ploughing natural habitats to be used as agricultural lands, excessive grazing and excessive usage of water and converting wetlands to agricultural lands as they make the impact of CC more complicated [15]. Figure 5.4 portrays green infrastructure such as urban agriculture, green walls and roofs.

5.1.6 *Human Health Policies*

It is essential to develop researches about climate variability risks on public health and map vulnerable populations and regions to monitor and evaluate the adverse health outcomes related to CC [24]. It is also vital to develop early warning (RWS)

Fig. 5.4 Green infrastructure. (Source: US EPA)

and emergency response plans and raise the awareness of medical staff and communities about CC-related diseases by creating educational programmes and workshops [25].

Other methods include developing programmes to increase access to safe water and improve sanitation to adapt to water diseases' risks. Additionally, it is significant to develop programmes to support local food production and sustainable food sources, support vulnerable groups (children and elderly people) to adapt to malnutrition due to CC impact on agriculture and impose air quality regulations to control emissions and pollution from activities including traffic and industry [26]. Moreover, Europe developed some research projects to deal with CC impacts on human health including quantifying projected impacts under 2 °C warming, developing strategies for cities to mitigate the CC impacts and establishing a European study of cohorts for air pollution effects [27].

5.1.7 Tourism Policies

The integrated measures in CC adaptation in the tourism sector include sustainable development planning and linking adaptation policies with policies in other sectors to avoid maladaptation [28]. It is imperative to generate new knowledge about current and future impacts of CC in cooperation with research centres and stakeholders to build adaptive experiences, enhance knowledge about the effects of CC on tourism regions and sectors, encourage strategies such as water recycling and water conservation to adapt to water shortage and create slope contouring and landscape to reduce the amount of snow required for skiing [29]. In Portugal, a policy was developed to compel tourist resorts to follow adaptation strategy through financial incentives, e.g. subsidies and tax credits, also induce building code in tourist resorts and use green architectural and construction solutions such as techniques and materials that have resistance to climate variables and protect shoreline form floods by sea wall and vegetate sand dunes [30].

Table 5.2 presents the global climate change adaptation policies in Africa, Europe, Asia, North America and Latin America. The table indicates that CC policies are adopted globally, but they differ in the effectiveness of the policy implementation on ground. Thus, adaptation decision framework represents a major element in active effectiveness of policies through moving adaptation from theory to practice. Also, it is important to make a decision in selecting appropriate adaptation measures and form an adaptation policy for the adaptation decision framework. Such framework is responsible for moving adaptation from theory to practice and managing CC risks uncertainty. The decision-making through adaptation framework goes through four stages: (1) assessment, (2) planning, (3) implementation and (4) monitoring and evaluation. They are fixed in different frameworks and represent the main stages of adaptation decision framework (ADF). The following section investigates the mapping of ADF.

Table 5.2 (a) Global climate change adaptation policies (1–2)

Climate change adaptation policies (CCAP)	Africa	Europe	Asia	North America	Latin America
Grey strategies as retrofitting, green as green infrastructure and soft as policies and regulation	✓ [31]	✓ [2]	✓ [32]	✓ [33]	✓ [33]
Incorporate adaptation into planning, design and operations	✓ [31]	✓ [34]	✓ [35]	✓ [33]	✓ [36]
Water management strategies	✓ [31]	✓ [37]	✓ [33]	✓ [38]	✓ [38]

✓ **Mark means that the policy is adopted in regional policies**
[ ] The red number between two red bracts is the reference

Table 5.2 (b) Global climate change adaptation policies (2–2)

Climate change adaptation policies (CCAP)	Africa	Europe	Asia	North America	Latin America
Diversification energy sources and improve energy efficiency	✓ [31]	✓ [2]	✓ [35]	✓ [38]	✓ [38]
Integrate CC adaptation into food security and (R&D) policies	✓ [39]	✓ [40]	✓ [33]	✓ [38]	✓ [33]
Capacity development to support mitigation and adaptation actions	✓ [40]	✓ [41]	✓ [32]	✓ [33]	✓ [36]
Enhance public health and impose air quality regulations	✓ [33]	✓ [41]	✓ [42], [43]	✓ [38]	✓ [38]
Ecosystem restoration and conservation	✓ [33]	✓ [2]	✓ [33]	✓ [38]	✓ [33]
Livelihood diversification and investments in renewable energy	✓ [31]	✓ [44]	✓ [45]	✓ [33]	✓ [38]
Integrated CCA in tourism plans and build adaptive capacity	✓ [29]	✓ [29]	✓ [29]	✓ [29]	✓ [29]
Total CCA policies	10	10	10	10	10

✓ **Mark means that the policy is adopted in regional policies**
[ ] The red number between two red bracts is the reference

5.2 Mapping of Adaptation Decision Framework

Adaptation decision framework, such as the IPCC framework shown in Fig. 5.5, aims at assessing climate change impacts and selecting appropriate adaptations decision through identifying problems and objectives of the framework. It then selects the analytical method to construct scenarios or projected ranges for key climatic variables and then tests the method's sensitivity to ensure credibility of model. The framework builds scenarios to investigate the impacts of climate change with respect to the environmental and socio-economic conditions occurring with and without climate change. Afterwards, it assesses the autonomous adaptation taken by individual for responding to climate change. Finally, it evaluates adaptation strategies by comparing between the performances of different strategies in meeting the objectives in order to select the best alternative adaptation strategies.

The following sections present in brief the seven steps of IPCC adaptation framework: (1) identify problems and objectives, (2) select suitable methods, (3) test method or sensitivity, (4) select scenarios, (5) assess the impacts, (6) assess adjustments and (7) evaluate adaptation strategies.

5.2.1 *Identify Problems and Objectives*

In this step, it is important to assess current and projected vulnerability separately for each sector or several sectors interactively in order to create accurate and specific results. Before the mapping of adaptation decision framework, the user must ensure the availability of needed data for studying. The data required for assessing the vulnerability depends on the goal of the adaptation strategy [46]. For instance, the data required for assessing heat island effect on energy consumption differ, if the goal is to estimate the impact on air quality and thermal comfort in cities. This step should assess the vulnerability through exposure, sensitivity and adaptive capacity of vulnerable people, regions and sectors, for example, the impact of climate change on water supply leads to drought but with high sensitivity and low adaptive vulnerability increase [47].

5.2.2 *Select Suitable Method*

It is essential to select analytical methods to construct scenarios or projected ranges for key climatic variables by using one or more of the four analytical methods. The first method is experimentation by evaluating causes and quantifying upper and lower limits of CC impacts through scientific experimentation. The second is forming an impact projection by evaluating climate model data and constructing climate scenarios [48]. Other methods include surveying the opinion of experts in the field

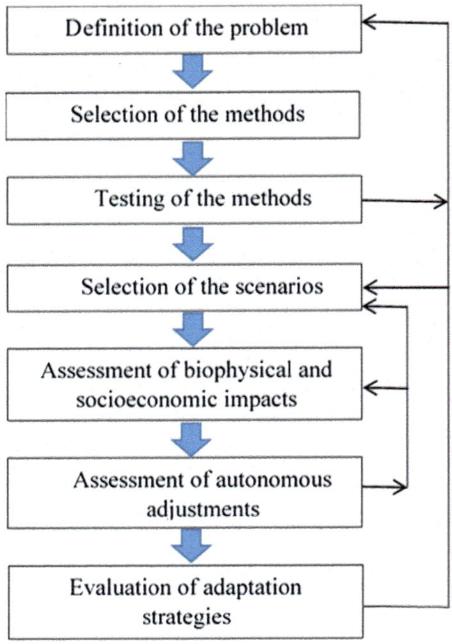

Fig. 5.5 IPCC adaptation framework. (Source: Inter-governmental panel on climate change)

or observing the interactions of climate and society in a region to predicate the future scenarios [46].

5.2.3 *Test Method or Sensitivity*

Test models are used to evaluate scenario models to estimate future impacts and compare between model predictions and current observations. It is developed to assess model errors and performance because the world cannot wait for the future to assess model performance to be able to early respond to CC effects to minimize losses [49].

5.2.4 *Select Scenarios*

Climate change impacts are described as the differences between environmental and socio-economic conditions expected to exist over the period of analysis in the absence of climate change and the occurrence of climate change. Such differences present the scenario regarding environmental conditions, which include changes in land use, air, water and soil pollution, in addition to socio-economic conditions which include economic growth and unemployment [46].

5.2.5 *Assess the Impacts*

The assessment of environmental and socio-economic impacts is carried out through assessing the impact in certain regions, activities or organisms that are sensitive to climate. The assessment uses environmental standards as a reference to measure changes, which occurs due to CC impacts. Assessment can also be conducted according to the impact on the economy or geographical extent of impacts [50].

5.2.6 *Assess Adjustments*

The assessment of the autonomous adaptation, taken by individual institutions, enterprises, and communities independently and spontaneously, is prepared to adapt to climate change impacts on an exposure unit such as activity, group and region and resource. Autonomous adaptation is the reactive or anticipatory adaptation to climate change, for example, purchasing of air conditioning units to adapt to the increasing high temperatures in summer or paying subsidies to community to adapt to economic losses [28, 51].

5.2.7 *Evaluate Adaptation Strategies*

The IPCC's framework that evaluates CCA strategies, to select and prioritise the best appropriate measures for coping with CC, is set through the following steps:

(a) Determine the goals and objectives of the adaptation strategies, for example: the promotion of sustainable development, and the reduction of vulnerability [52].
(b) Specify and assess climatic impacts and their importance according to the resulting losses in order to select the appropriate adaptation strategy [53].
(c) Identify the adaptation options, through determining adaptive responses to climate change from six types of strategies, preventing losses by anticipatory adaptation with impacts and tolerating impacts which have little or no negative impact on the long term [46].
(d) Examine the constraints; examine the effects of the constraints such as knowledge, social and technology barriers on adaptation options [54].
(e) Quantify measures and formulate alternative strategies; evaluate the performance of adaptation measure to achieve the specified objectives in the previous steps under different scenarios through simulations models or expert judgement [55].
(f) Weight the objectives and evaluate trade-offs, and then compare between the performances of different strategies and integrate them with the national sustainable development plans [52].

Table 5.3 Comparison between IPCC and APF adaption framework

Criteria of comparison	Assessing climate change impacts and adaptations guidelines (IPCC)	Adaptation Policy Framework (APF)
Generation of adaptation policy	First generation 1995	Second generation 2004
Developer	Intergovernmental Panel on Climate Change (IPCC)	United Nations Development Programmes (UNDP)
Scope	All regions and sectors	All sectors, all regions particularly developing countries
Adaptation methodology	Based on a seven-step approach 1. Definition of the problem 2. Selection of the methods 3. Testing of the methods 4. Selection of the scenarios 5. Assessment of biophysical and socio-economic impacts 6. Assessment of autonomous adjustments 7. Evaluation of adaptation strategies	Based on a five-step approach 1. Defining project scope and design 2. Assessing vulnerability under current climate 3. Characterizing future climate-related risks 4. Developing an adaptation strategy 5. Continuing the adaptation process
Focus points	Assessing the impacts of potential climate change and evaluating appropriate adaptations	Better centred on adaptation
Key input	Depends on the objectives, available information and cost-benefit analysis and economic models	Depends on the particular application available information, vulnerability mapping and dynamic simulation of sustainable livelihoods and multi stakeholders and cost-effectiveness analysis
Key output	Focuses on minimizing the effects of climate change	Focuses on consider adaptation in the development context to increased adaptive capacity
Time frames of measures	Longer-term adaptation measures	Both short- and longer-term adaptation measures
Community participation	Survey the opinions of experts only	Focuses on the involvement of stakeholders and community at all stages
Flexibility of adaptation framework	Prescriptive	Less prescriptive
Addressing of regional adaptation strategy	Comparable results across regions and sectors	Addressing regional and transboundary issues in adaptation

(g) Recommend adaptation measures; recommend best adaptation measures to policy advisors and decision-makers, and recommend adaptation measures derived from previous evaluation process [46].

Nonetheless, Table 5.3 presents the comparison between the first generation of adaption framework (IPCC) and the second generation (APF). The first generation appeared in 1995 by the Intergovernmental Panel on Climate Change (IPCC) to assess the impacts of potential CC and evaluate appropriate adaptations. The scope of the framework includes all regions and sectors, and the adaptation methodology was based on a seven-step approach. Key input depends on the objectives, available information, cost-benefit analysis and economic models to minimize the effects of CC through long-term adaptation measures. Also, a survey was implemented to gather the opinions of experts only in weak community participation. There are other frameworks such as Adaptation Policy Framework (APF) that represents the second generation of adaption framework that seeks to also support adaptation processes. In addition, it seeks to consider adaptation in the development context in order to increase adaptive capacity and focus on the involvement of stakeholders at variance with the first generation that concentrated on experts' opinion only. Therefore, APF is preferable in developing countries due to their aspiration for the enhancement of sustainable development. Also, it takes the advantage of enormous population by supporting autonomous adaptation by individuals and NGOs to alleviate poverty and CC together. Adaptation methodology is based on five steps: (a) defining project scope and design, (b) assessing vulnerability under current climate, (c) characterizing future climate-related risks, (d) developing an adaptation strategy and (e) continuing the adaptation process. The framework addresses regional and transboundary issues in adaptation. Therefore, in light of the above sections,

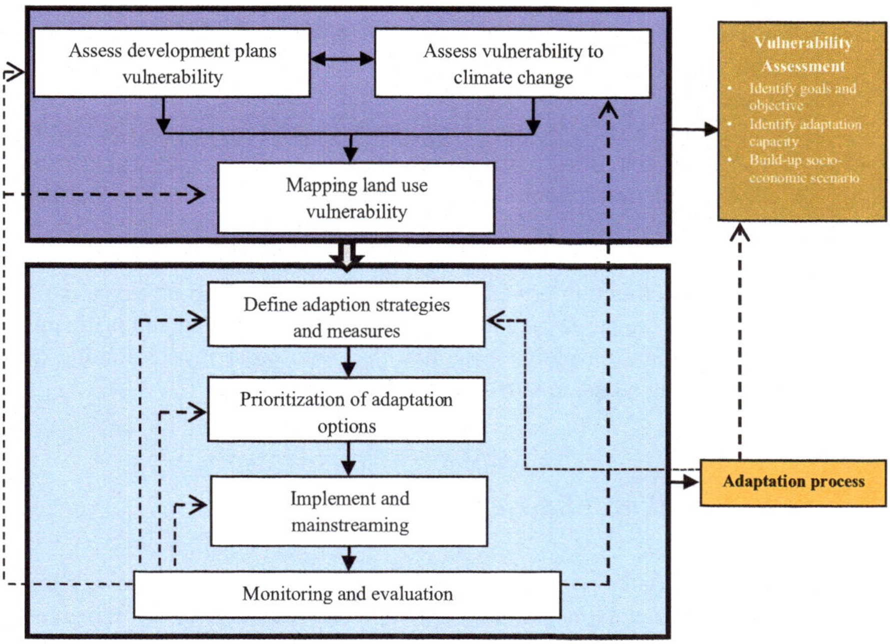

Fig. 5.6 Sustainable climate change adaptation framework. (Source: Developed by authors)

applying sustainable climate change adaptation (CCA) framework is a necessity step. Fig. 5.6 outlines the steps involved in brief. It is important to note that CC adaptation in Africa faces barriers that reduce its effectiveness and hinder CCA which makes Africa the most vulnerable continent to CC.

Therefore, the next section investigates CCA barriers, discuss the impact on adaption and recommend strategies to overcome obstacles as shown in Table 5.4.

5.3 Barriers to Climate Change Adaptation Policies in Africa

Barriers are defined as obstacles that can be overcome with concerted effort [56]. Developing countries are at risk as they are more reliant on environment resources and more vulnerable to coastal and water resource change and have less financial, technical and institutional capacity to adapt [57]. Africa is a particularly vulnerable continent due to the enormous impact of CC, low existing adaptation capacity, increased poverty rates and rapid population growth in absence of active development plans.

5.3.1 Financial Barriers

Financial barriers are largely related to budget deficits in many economies and the increased of poverty rates. For example, adapting with sea level rise is technologically possible, but large-scale engineering measures for coastal protection are beyond the reach of many governments due its high costs [54].

Lacking scientific capacity in developing countries to assess risks, vulnerabilities and associated uncertainties needs to be strengthened in order to create active adaptation policies. Adaptation capacity faces difficulties to formulate costs and bankable projects and failure in creating projects and programmes derived from strategies and plans. Thus, those countries face difficulties in meeting international funding requirements and in attracting private sector investment [58]. Micro-financing and social welfare grants are based on a long-term sustainable basis and mainstreaming adaptation into national development policy that will assist in overcoming constraints and enhancing adaptation to climate change [59].

5.3.2 Social-Cultural Barriers

The cultural background of people greatly influences the way they perceive climate change and act with adaptation strategies. Also, it can constitute one of the greatest barriers to the implementation of adaptation for households; for example, cultural practices prevent migration of female farmers in northern Burkina Faso to adapt to

Table 5.4 Climate change adaptation barriers in Africa

Barriers		Results	Recommendations
Financial barriers	Budget deficits	Failure in moving adaptation from theory to practice Increase vulnerability Low financial scientific budget	Micro-financing and social welfare grants Transition to a sustainable green economy
	High poverty rates		
	Absence of green investment		
Social-cultural barriers	Illiteracy and wrong cultural legacies	Influence the way communities perceive CC	Involvement of stakeholders Raise community awareness
Information barriers	Lack scientific research support	Low scientific capacity Limit the ability to create adaptation plan and make decision Difficulties in meeting international funding requirements	Improve scientific research and support researcher
	Low research financial funding		
Technological barriers	Low technical expertise in climate modelling	Low technical expertise in climate modelling Limit experimental science research	Support scientific research Support international scientific partnership and training
	Technological backwardness		
Development barriers	Lack of development infrastructure and Poverty	Limit response to CC adaptation projects	Sustainable development Enhance community resilience
	Lack eco-markets		

droughts [60]. On the other side, Samoa communities in the south pacific rely on livelihood diversification and financial remittances through extended family networks to cope with storm damage [28]. Thus, governments or organization should focus on the involvement of stakeholders at all stages of adaptation because communities differently response to risks.

5.3.3 *Information Barriers*

Some people in developing countries are concerned with violent conflict, disease and hunger which may overshadow considerations about the impacts of CC and its adaptation [28]. Thus, institutions should focus on CC impact on conflict, disease and hunger to attract the world's attention to the seriousness of CC. The lack of information on CC limits the community interaction with adaptation strategies and limits the ability of governments in creating an effective adaptation plan and making decision. Thus, governments should raise community awareness about CC, improve scientific research and support researcher efforts by providing financial fund which will help governments in enhancing CC adaptation.

5.3.4 Technological Barriers

Low technology availability prevents the development of advanced materials and early warning systems that are critical to climate adaptations due to low financial scientific budget and minimum training and knowledge of the researchers. The Africa climate report commissioned by the United Kingdom identified low technical expertise in climate modelling as challenges to CC research in Africa. Lack of technology is also associated with inadequate funding on research. Therefore, this demonstrates a close link between technological barriers and institutional barriers [60].

5.3.5 Development Barriers

The lack of infrastructure development and lack of eco-markets will limit the access of consumers to adaptation techniques. Therefore, producers will not be able to recover what they paid or repay their loans. Generic indicators, including factors such as education, income and health, play a role in the response of groups to adaptation process. The solution is in sustainable development that focuses on economic growth, social progress, environmental conservation and adaptation to CC. Till present, Africa is the most continent affected by CC risks and also the most vulnerable to CC as demonstrated in Chap. 2. Additionally, the slow rate of development and poverty redoubled the effects of CC, hence played a major role in increasing such problems. Thus, most of African countries are vulnerable to CC such as Egypt. According to a study produced by the Met Office, they considered the impact of a 1.00 m SLR for 84 developing countries, Egypt was ranked the second highest with respect to the coastal population affected, third highest for coastal GDP affected and fifth highest for proportion of urban areas affected. The next chapter focuses on investigating CC impacts on Egypt and also measures and policies taken to adapt in Egypt.

5.4 Conclusion

This chapter investigates climate change adaptation (CCA) policies in urban areas, infrastructure, water resources, energy, green infrastructure, public health and tourism sector. It also identifies their role in buildings' retrofitting and cities' upgrading. Investigation indicates that global climate change adaptation strategies on the built environment are divided into three strategies: grey strategies, green strategies and soft strategies. Grey strategies include passive cooling to mitigate global warming, while green strategies include green infrastructure to reduce air pollutants and storm water runoff, and soft strategies include plans related to decreasing climate change risks (CCR) through considering CCR into land use regulation and building codes to increase energy efficiency and improve the durability of buildings structure to floods.

Policies are adopted globally, but there is difference in the effectiveness of policy implementation on the ground. The chapter identifies adaptation decision framework that is responsible for the formation and prioritisation of policies in order to be integrated into sectors plans. It also focuses on national climate actions, moving adaptation from theory to practice and selecting appropriate adaptation measure which goes through adaptation decision framework. Investigation indicates that there are different adaptation frameworks. Every framework seeks to assess vulnerability and prioritise solutions while taking into consideration addressing the shortcomings from the perspective of the developer.

Mapping adaptation framework goes through five stages; assessment, planning, implementation, monitoring and evaluation, which are fixed in different CCA frameworks.

Finally, the chapter determines the challenges and barriers adaptation process faces and how to overcome them for better adaptation practice. It is indicated that Africa is the most vulnerable continent due to enormous CCR, low adaptation capacity, the spread of ignorance, poverty and rapid population growth in absence of active development that represent barriers to adaptation. However, this can be addressed through the transition to a sustainable green economy, enhancement of community resilience and building scientific capacity to fill the information gap.

Africa is the most continent affected till now by CC and also the most vulnerable continent to CC as demonstrated in Chap. 2 due to slow rate of development and poverty that redoubles the effects of CC. Therefore, most of African countries are vulnerable to CC including Egypt. According to study produced by the Met Office, they considered the impact of a 1 m sea level rise (SLR) for 84 developing countries; Egypt was ranked the second highest with respect to the coastal population affected, third highest for coastal GDP affected and fifth highest for proportion of urban areas affected.

The next chapter will focus on investigating CC impacts on developing countries and measures and policies that are taken into consideration to adapt to CC risks.

References

1. Susanne Menzel and Marco Pütz, "Climate Adaptation Policies and Governance in Alpine Countries", Swiss Federal Institute for Forest, Snow and Landscape Research (WSL), Bern, Switzerland, July 2013.
2. "Urban adaptation to climate change in Europe Challenges and opportunities for cities", supportive national and European policies European Environment Agency, 2012.
3. Blair Feltmate and Jason Thistlethwaite, "Climate Change Adaptation: A Priorities Plan for Canada", The Climate Change Adaptation Project (Canada), 2012.
4. The International Bank for Reconstruction and Development, "Guide to Climate Change Adaptation in Cities", The World Bank, 2011.

5. "Climate Adaptation Strategy" City of Surrey, November 2013.
6. "City of North Vancouver Climate Change Adaptation Plan" The City of North Vancouver, October 2013.
7. "Climate Change Adaptation Strategy" City of Vancouver and Greenest city 2020, [Online]. Available at: http://vancouver.ca/files/cov/Vancouver-Climate-Change-Adaptation-Strategy-2012-11-07.pdf. (Accessed: August 21, 2015).
8. Mark Snow and Deo Prasad, "Climate Change Adaptation for Building Designers: An Introduction", Environment Design Guide, February 2011.
9. Michael Bruch et al, "Power Blackout Risks Risk Management Options Emerging Risk Initiative", CRO Forum, November 2011.
10. "Effective Emergency Management: Making Improvements for Communities and People with Disabilities", National Council on Disability, 2009.
11. Gyda Grendstad et al., "Adaptation to climate change", CEDR's Secretariat General, 2012.
12. Giuseppe Inturri and Matteo Ignaccolo, "GRaBS Policy Guidelines Summary adapting transport systems to climate change", European Union (EU), April 2011.
13. Harris Switzman, "Mainstreaming Climate Change Adaptation in Canadian Water Resource Management", A.D. Latornell Conservation Symposium, Toronto, Canada, November 2012.
14. Ger Bergkamp, Brett Orlando and Ian Burton, "Change Adaptation of water resources management to climate change", International Union for Conservation of Nature and Natural Resources, 2003.
15. Ministry of Environment and Urbanization, "Turkey's National Climate Change Adaptation Strategy and Action Plan", Ministry of Environment and Urbanization, November 2011.
16. "Climate Ready Water Utilities Adaptation Strategies Guide for Water Utilities", US Environmental Protection Agency, January 2012.
17. Kundzewicz, Z.W. et al., "Freshwater resources and their management", in *Climate Change 2007: Impacts, Adaptation and Vulnerability*, Cambridge, UK, Cambridge University Press, pp. 173–210, 2007.
18. Division of Environment, "National Climate Change Strategy", United Republic of Tanzania Vice President's Office, Tanzania, 2012.
19. United Nations Development Programme, "National Climate Change Adaptation Strategy", United Nations, 2012.
20. "California climate adaptation strategy", California Natural Resources Agency (CNRA), 2009.
21. Asti Methodology and Data Collection Standards, "Practitioner's Guide for National and Regional Focal Points" 2002. [Online]. Available at: www.asti.cgiar.org/sites/default/files/pdf/methodology/ASTI-Practioners-Guide-partII-General-definitions.pdf. (Accessed: August 21, 2015).
22. Universidad de Politécnica de Madrid, "Adaptation to Climate Change in the Agricultural Sector, Energy & Environment", European Commission, December 2007.
23. W. F. Mwase et al., "Assessment of Agricultural Sector Policies and Climate Change in Malawi – The Nexus between Climate Change Related Policies", African Climate Policy Centre, UN Economic Commission for Africa, May 2013.
24. Kristie L. Ebi, Nancy D. Lewis and Carlos F. Corvalán, "Climate variability and change and their health effects in small island states: information for adaptation planning in the health sector", US National Library of Medicine, National Institutes of Health, 2006.
25. Tom Mitchell and Thomas Tanner, "Adapting to climate change Challenges and opportunities for the development community", Institute of Development Studies, Tearfund, 2006.
26. Kristie Ebi, "Protecting Health from Climate Change Vulnerability and Adaptation Assessment", World Health Day, 2008.
27. "Adaptation to climate change impacts on human, animal and plant health, An EU Strategy on adaptation to climate change", European Commission, 2013.
28. M. L. Parry, et al., Climate Change 2007: Impacts, Adaptation and Vulnerability, Cambridge University Press, Cambridge, United Kingdom, 2007.

29. Simpson, et al., "Climate Change Adaptation and Mitigation in the Tourism Sector: Frameworks, Tools and Practices", United Nations Environment Programme, 2008.
30. Alain Dupeyras, et al., "Climate change and Tourism Policy in OECD Countries, Tourism Committee of the OECD", United Nations Environment Programme (UNEP), 2010.
31. "Climate Change Action Plan 2011–2015", African Development Bank Group.
32. "Cities at Risk: Developing Adaptive Capacity for Climate Change in Asia's Coastal Mega Cities" [Online]. Available at: https://www.eastwestcenter.org/fileadmin/stored/pics/Cities%20at%20Risk%20prospectus_final.pdf. (Accessed: March 29, 2016).
33. The Intergovernmental Panel on Climate Change (IPCC), "Climate Change 2014: Impacts, Adaptation and Vulnerability", Cambridge University Press, Cambridge, UK, 2014.
34. "Climate Change Adaptation and Disaster Risk Reduction in Europe - A Review of Risk Governance" Council of Europe EUR–OPA Major Hazards Agreement, UNISDR Regional Office Europe and Central Asia, June 2011.
35. the Asian Development Bank (ADB) and the Government of Germany, [Online]. Available at: http://cdia.asia/projects/page/3/. (Accessed: December 14, 2015).
36. "Latin American and Caribbean Cities Take on Climate Change", The World Bank, March 2013.
37. The European Commission environment, [Online]. Available at: http://climate-adapt.eea.europa.eu/water-management. (Accessed: April 23, 2015).
38. Rachel M. Gregg et al., "The State of Marine and Coastal Adaptation in North America: A Synthesis of Emerging Ideas", EcoAdapt, January 2011.
39. Miriam Joshua, Abdulai Jalloh and Sepo Hachigonta, "Review of Research and Policies for Climate Change Adaptation in Urban Areas in Southern Africa", International Development Research Centre (IDRC), July 2014.
40. The International Bank for Reconstruction and Development, "The status of Disaster Risk Reduction in Sub-Saharan Africa", The World Bank, November 2010.
41. "The EU Strategy on adaptation to climate change", European Commission, April 2013.
42. "Climate Risk Management Approach to Disaster Reduction and Adaptation to Climate Change, UNDP Expert Group Meeting" in *Integrating Disaster Reduction with Adaptation to Climate Change Havana*, June 17–19, 2002.
43. James D. Ford et al, "The status of climate change adaptation in Africa and Asia", *Springer,* vol. 15, no. 5, pp. 801–814, June 2014.
44. Gabriella Camarsa et al., "LIFE and Climate change adaptation", Office of the European Union, Luxembourg, 2015.
45. "ICLEI – Local Governments for Sustainability South Asia" [Online]. Available at: http://southasia.iclei.org/get-involved/contact-us.html. (Accessed: April 25, 2016).
46. Timothy Carter et al., "Technical Guidelines for Assessing Climate Change Impacts and Adaptations", National Institute for Environmental Studies, University College London, 1994.
47. ICF International, "Climate Change Vulnerability Assessment, Risk Assessment, and Adaptation Approaches", Federal Highway Administration Office of Environment and Planning, 2009.
48. Roger Neville Jones, "An Environmental Risk Assessment/Management Framework for Climate Change Impact Assessments", *Springer Nature,* vol. 23, no. 2–3, pp. 197–230, March 2001.
49. GPWayne, "How reliable are climate models?" 7 July 2015. [Online]. Available at: www.skepticalscience.com/climate-models.htm. (Accessed: August 18, 2015).
50. "Public Consultation on Hong Kong's Climate Change Strategy and Action Agenda" Environmental Protection Department, 2010.
51. "Types of Adaptation Measures", The Energy and Resources Institute, [Online]. Available at: know.climateofconcern.org/index.php?option=com_content&task=article&id=148#. (Accessed: August 19, 2015).
52. Mohan Munasinghe and Rob Swart, "Primer on Climate Change and Sustainable Development: Facts, Policy Analysis, and Applications", Cambridge University Press, Cambridge, UK, 2005.

53. Karen O'Brien, "Developing Strategies for Climate Change: The UNEP Country Studies on Climate Change Impacts and Adaptations Assessment", Center for International Climate and Environmental Research, Oslo, July 2000.
54. Adger, W.N. et al., "Ch.17: Assessment of adaptation practices, options, constraints and capacity" in *Climate Change 2007: Impacts, Adaptation and Vulnerability*, Cambridge University Press, Cambridge, UK, pp. 717–743, 2007.
55. Jan F. Feenstra et al., "Handbook on Methods for Climate Change Impact Assessment and Adaptation Strategies Version 2.0", United Nations Environment Programme, October 1998.
56. Susanne C. Mosera and Julia A. Ekstromc, "A framework to diagnose barriers to climate change adaptation", U.S. National Library of Medicine, 2010.
57. Tom Mitchell, Thomas Tanner and Emily Wilkinson, "Overcoming the barriers: Mainstreaming climate change adaptation in developing countries", TEARUND, TEARUND, 2006.
58. Michiel Schaeffer et al., "Africa's Adaptation Gap(technical report)", United Nations Environment Programme, 2013.
59. Michel Boko et al., "Ch.9: Africa, IPCC Fourth Assessment Report: Climate Change 2007" in *Climate Change 2007: Impacts, Adaptation and Vulnerability*, Cambridge University Press, Cambridge, UK, pp. 433–467, 2007.
60. Philip Antwi, Andrew J. Dougill and Lindsay C. Stringer, " Barriers to climate change adaptation in subSaharan Africa: Evidence from northeast Ghana & systematic literature review", Sustainability Research Institute, The University of Leeds UK, October 2013.

Chapter 6
Risks, Policies, Measures and Actions for Climate Change Adaptation in Developing Countries

Abstract This chapter investigates climate change adaptation in developing countries. In line with the fact that Africa is the most vulnerable continent, therefore Egypt, as one of the developing countries, is selected as a case study since it is one of the most exposed countries to climate change risks. Climate change impacts, adaptation policies and adaptation measures in developing countries were reviewed, presented and discussed. The chapter is laid out in three main folds: the first part discusses climate change risks on urban and coastal areas, infrastructure and related economic sectors including agriculture, public health, ecosystem, fishing, tourism and politics; and the second and third folds highlight the adaptation policies and measures adopted by government authorities and civil society. These policies, strategies and measures include building up the scientific capacity, developing adaptation strategy to increase energy and water efficiency, enhancing land use management, improving public health and retrofitting vulnerable buildings and infrastructure. It is important to note that this chapter examines how the action plans address the risk of climate change in order to cope with the impacts or consequences of climate change in the short or medium term.

6.1 Introduction

Climate change poses a fundamental threat to all sectors, regions and societies. Low development rates, illiteracy and conflicts make developing countries the most vulnerable to risks. Negative consequences of climate change (CC) are represented in many parameters: sea level rise, water scarcity, food insecurity and pressure on human health, ecosystems and the national economy. All CC risks, whether directly or indirectly, affect cities and their infrastructure, which are already suffering from induced human pressure and the absence of sustainable development implementation.

The impact of climate change in developing countries will be discussed. Also, the adaptation policies developed and adaptation measures taken to offset drastic effects of climate change will be reviewed and conversed. These adaptation policies will be examined in depth in one of the developing countries in Africa. Egypt is

M. M. Aboulnaga et al., *Urban Climate Change Adaptation in Developing Countries*, https://doi.org/10.1007/978-3-030-05405-2_6

considered as a reference case, since it is considered one of the potential vulnerable developing countries to CC impacts, despite the fact that its greenhouse gas (GHG) emissions form about 1% of the total global GHG emissions.

6.2 Climate Change Impact on Developing Countries

Current impacts of climate change are creating a major challenge to developing countries. As indicated earlier in the previous chapters, climate change (CC) is not only something that will occur in the future; nevertheless, it has significantly occurred severely over the last 30 years. Over the last century, the average temperature around the globe has increased by 0.75 °C. According to IPCC 2007, 11 of the 12 hottest years occurred alone during the last 12 years. Also, the rainfall patterns have also changed, thus leading to increasing the number of droughts, heat waves, extreme rainfall events and more intense cyclone activity in different parts of the world as shown in Fig. 6.1 [1]. These extreme events, particularly, affect developing countries, which are more vulnerable to climate change as illustrated in Fig. 6.2.

Additionally, during August and September 2018, the typhoon 'JEBI' that struck Japan was immense and destructive, and typhoon Mangkhut that hit Hong Kong, China and the Philippines was also seriously severe and devastating. Both led to the destruction of many parts of the cities and their infrastructure (Osaka and Kyoto in Japan and Manila in the Philippines) in these countries. On the western side of the Atlantic, the cyclone 'Florence' that hit North and South Carolina, USA, was also destructive, and many lives were lost, besides a large number of injures [2]. Figures 6.3 and 6.4 depict some of these devastations and destructions in Japan; the Philippines; Hong Kong, China; and North Carolina, USA.

According to IPCC 2007, developing countries will have a disproportional negative impact caused by CC due to the fact that these countries have limited capacity to withstand and cope with the adverse effects of CC. This is basically due to the

(a) (b)

Fig. 6.1 Climate change impacts in developing countries. (**a**) Snow covering the Sahara desert in Algeria. (https://www.express.co.uk/news/world/901733/Sahara-Desert-snow-Ain-Sefra-Algeria-pictures-photos). (**b**) Flooded area in Kisumu in Kenya. (https://qz.com/africa/1271077/floods-in-kenya-somalia-displace-hundreds-of-thousands/)

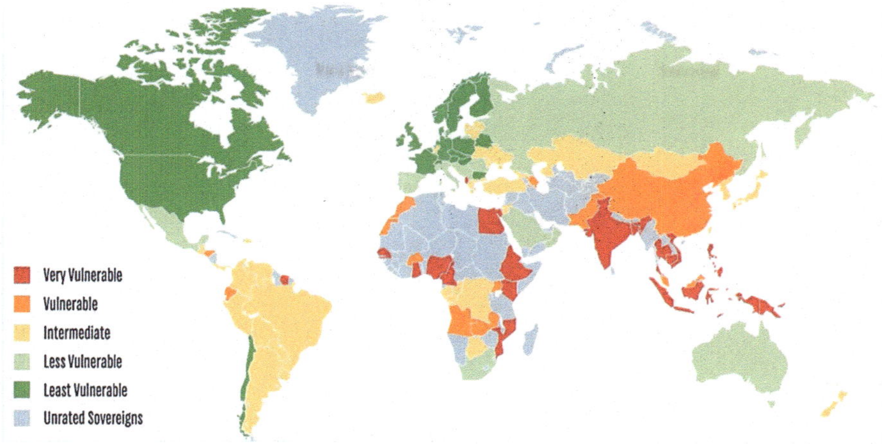

Fig. 6.2 Developing economies are likely to be most impacted by global warming. (Source: Wade & Jennings [3])

Fig. 6.3 Severe events in many Asian countries – August–September 2018. (**a**) Flooded areas in Kyoto, Japan. (Image source: https://www.bbc.com/news/world-asia-44749847). (**b**) Destructed areas in Osaka, Japan. (Image source: https://www.theguardian.com/world/gallery/2018/sep/05/typhoon-jebi-batters-japan-in-pictures). (**c**) Destructed and flooded areas in the Philippines. (Image source: https://www.telegraph.co.uk/news/2018/09/15/super-typhoonmangkhut-kills-least-eight-smashes-philippines/). (**d**) Destructed and flooded areas in Hong Kong. (Image source: https://www.vox.com/2018/9/17/17869462/typhoon-mangkhut-philippines-hong-kong-china)

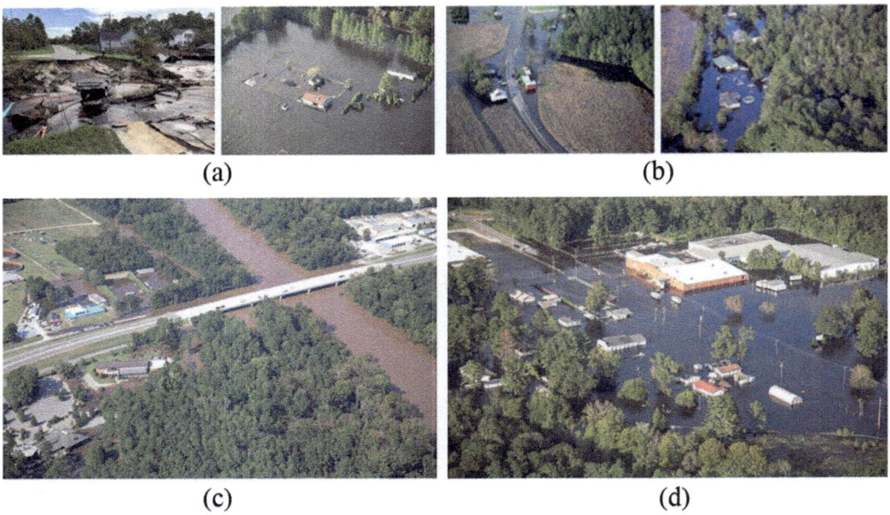

Fig. 6.4 Severe storms and flooded areas in North Carolina, USA – September 2018. (a) Destructed and homes flooded with water Eastern, North Carolina. (Image source: North Carolina National Guard). (b) Homes surrounded by flooded flowing out of Cape Fear River, eastern N.C. (Image source: North Carolina National Guard). (c) Flooded area nearby Cape Fear River – East of North Carolina. (Image source: North Carolina National Guard). (d) Home and Business flooded nearby Cape Fear River – East of North Carolina. (Image source: North Carolina National Guard)

fact of lacking both human and financial resources needed for effective adaptation measures and actions. Also, developing countries are the most vulnerable to CC risks due to the sensitivity of their fragile environments as minor changes in the climate led to the occurrence of huge environmental changes such as rapid desertification. In addition, climate change is a sensitive topic in developing countries since it affects their economies that primarily depend on agriculture and forestry [1]. Figure 6.5 plots the overall vulnerability ranking against the gross domestic product (GDP) per capita for each country, which indicates that poor countries are more vulnerable to CC.

Examples of CC impacts that occurred in developing countries are demonstrated in Southern Africa due to the fact that it has relatively longer dry season and more uncertain rainfall, which led to a decrease in agricultural production and forced people to adapt through switching crops. In Africa, lower lake levels have been observed in Zimbabwe, Zambia and Malawi, whereas in Latin America and Asia, glaciers are observed to be melting due to global warming; and mountain snow packs are disappearing. In addition, the risk of glacial lake outburst floods (GLOF) is increasing in the Andes and the Himalaya. The melting of glaciers led to a reduction in the stream flows in some rivers especially during dry seasons, as these glaciers are often considered as important freshwater sources. This is demonstrated in the Andes, where smaller glaciers have already disappeared or are expected to in the near future. In addition, in many tropical countries, nearly all the annual rainfalls

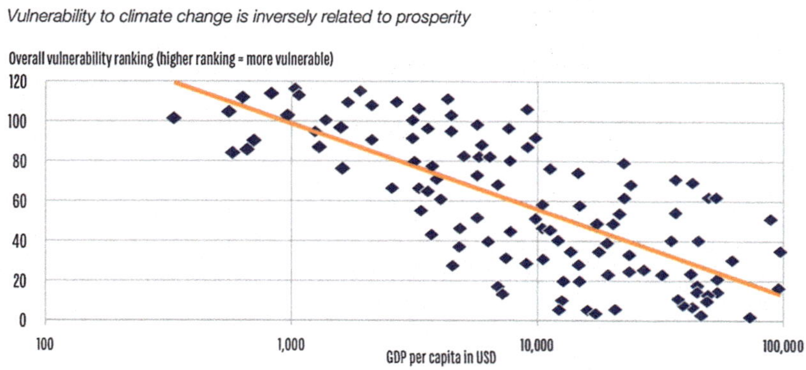

Fig. 6.5 Higher vulnerability trend among poorer countries. (Source: Wade & Jennings [3])

occur within specific few months. If the monsoon fails, no rainfall is present for almost a year until the next rain comes; thus, this leads to agriculture losses and water resources reduction [1].

6.2.1 Future Impacts of Climate Change on Developing Countries

According to CC high-emission scenarios, temperature is expected to rise between 1 °C and 6 °C in the next century. The amount of rainfall is expected to decrease in most of the sub-tropical regions where many developing countries are located, while rainfall will increase around the tropics and in the northern part of the Northern Hemisphere (Canada, Russia and Northern Europe). Regarding developing countries, future scenarios predict many consequences that demonstrate climate change. In fact, most of the developing countries are in the sub-tropics, where natural climate variability is high [1]. Furthermore, many developing nations are located in low-latitude countries, and it is predicted that 80% of negative effects of climate change will be concentrated in these areas [3]. Nonetheless, the following points highlight the main future climate change impacts in developing countries:

- Temperature is already very high in many developing countries and is expected to increase, which will severely affect agriculture as some crops will not tolerate much warmer weather [1].
- Changing rainfall patterns also threaten agriculture in Africa (specifically in the Sahel, East Africa and Southern Africa), as by 2020; in some developing countries, it will reduce the presence of rain-fed agriculture by as much as 50% [4]. Also, in sub-Saharan Africa, food security will be a major challenge due to droughts and shifts in rainfall [3].

- Receding glaciers create a risk in South Asia and Western South America as it may convert some of the world's mightiest rivers into seasonal flows by 2030, which will also severely affect the agriculture activity [3]. Furthermore, by 2035, the agriculture productivity in India is expected to decrease by 30–40% as the Ganga, the Brahmaputra and the Ganges basins will experience a reduced water flow due to loss of snowmelt from the Himalayas [3].
- Flooding will increase in South Asia specifically in cities such as Kolkata and Mumbai along with facing high temperatures and intense cyclones [3].
- Sea level rise (of 40 cm) by the end of the century would result in about 35 million refugees in the Bangladesh delta [4]. Furthermore, Southeast Asia, specifically Vietnam's Mekong Delta, which is responsible for the production of the country's rice, is vulnerable to rising sea levels [3].
- Storms and coral bleaching is expected to destroy fishing grounds and farming systems, specifically in the world's Coral Triangle, which is located at several developing countries in Southeast Asia including Indonesia. This area is responsible for supplying more than 50% of fish nutrients for the total world consumption which will severely impact fishing industry and ecosystem [5].

Box 6.1 Climate Change Challenges and Response for Natural Heritage in Developing Countries (*Presented at International Academy Cottbus: Understanding Heritage (ISAC 2015) Hosted by Brandenburgische Technische Universität (BTU) Cottbus-Germany December 12, 2015*)
Climate change has enormous negative effect on natural heritage. The increased intensity and frequency of extreme weather events, such as increase global temperature and SLR over the past decades and other negative impacts of CC on natural heritage, ring an alarm bell to announce that worldwide ecosystem is under threat. Climate change has a negative impact on terrestrial and marine ecosystems by altering the pattern of the ecosystem, causing widespread species extinction, migration and behaviour changes. Global natural heritage already suffers from CC, and in developing countries, situation of natural heritage is not better but worst due to human-induced pressures on the environment and the absence of sustainable development.

6.2.2 *What Is the Impact of Climate Change on Egypt?*

Has climate change already manifested its risks in developing countries including Egypt? Indeed yes! Egypt, a developing country, as a reference case demonstrates the negative consequences of CC represented in sea level rise and global warming which result in water scarcity, food insecurity and deteriorating human health and ecosystems which increase the pressure on the national economy. Climate change impacts on the built environment and related economic sectors, including urban areas, coastal zones, infrastructure, water and agriculture as well as public health, ecosystems, fishing and tourism, will be presented and discussed.

6.2.2.1 Climate Change Impact on Urban Areas

Urban areas in developing countries such as Egypt have been exposed to various types of climate change severe events and its impacts. One of which is the rise in temperature in cities due to urban heat island phenomena which causes air pollution in such cities. Higher temperatures increase ozone pollution in ground level of the earth, which exacerbates diseases such as asthma and breathing difficulties even in healthy individuals. In Egypt, a heat wave had hit Cairo from August 9 till 18, 2015, with high air temperature record at 46 °C (Fig. 6.6). This extraordinary heat wave resulted in hundreds of people suffered from heat stress at hospitals, and 106 elderly were dead [6].

Increased air pollution and heat stress affect public health, especially children, the elderly and patients with respiratory diseases, and cause mortalities [7]. Egypt has air quality problems resulted from air pollution. Statistics shows that there are 20,000 people who died annually due to diseases caused by air pollution. In Egypt, air pollution costs the economy 1–3% of its GDP [8]. Global warming effects have been increasing in Egypt since the 1960s, which led to an augment in the frequency of warm nights accompanied by less cool nights between 1960 and 2003. Also, an increase in average summer temperatures was recorded due to global warming. It is anticipated that air temperature is expected to increase over Egypt between 3 °C and 3.5 °C by 2100 [9]. This will consequently increase the cooling demands in summer months and, thus, elevate the demand on energy and electricity supply to buildings. Figure 6.7 shows the UHIE in downtown Cairo. In Fig. 6.7, the red areas illustrate the increased effect of urban heat island (UHI) effect in downtown Cairo which has led to average welfare loss estimated to be 2.5 billion EGP by 2030 and 14.7 billion EGP by 2060 as shown in Table 6.1.

(a) (b)

Fig. 6.6 Heat wave hit Cairo, Egypt, in August 2015. (**a**) Elderly walking under the shade, Cairo Street. (**b**) Children in a foundation, Al-Azhar Park. (Source: www.google.com)

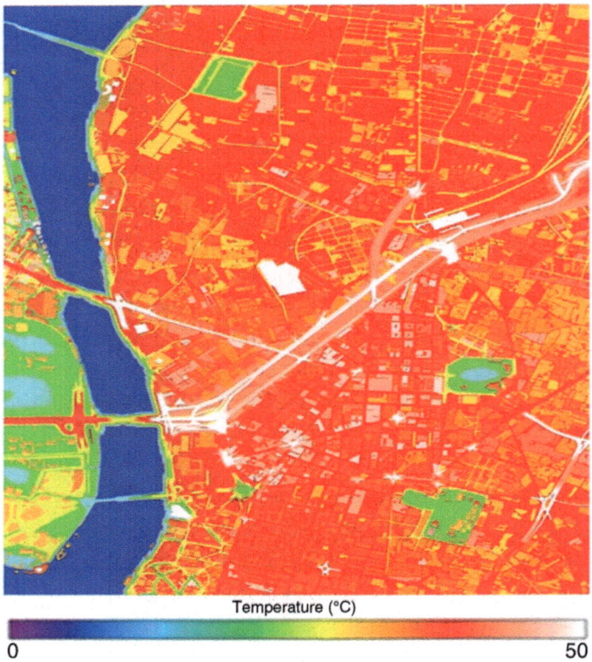

Fig. 6.7 Heat island effect in downtown of Cairo. (Source: method design)

Table 6.1 Estimated annual welfare loss from heat stress in Cairo (billions EGP)

Socioeconomic scenario/ climate change scenario	2030			2060		
	0.9/−4%	0.9/0	1.0/−5	2.0/−10	1.9/0	2.2/−10
Low population/high GDP	3.3	3.3	3.7	25.0	23.7	28.9
High population/low GDP	2.4	2.4	2.7	14.2	13.5	16.4

Source: UNDP
The first number in the climate change scenarios cells is the estimated °C increase in temperature. The second number is the percentage change in precipitation

6.2.2.2 Climate Change Impact on Coastal Zones

The long Mediterranean coast (1200 km) represents the northern coastal zone of Egypt and Nile Delta. These areas are exposed to inundation and loss of land estimated to 704 kilometre square by 2025. Also, due to sea level rise (SLR), a study predicts that 50 cm rise in sea level would threaten two million people in Alexandria alone.

Middle SLR scenario projects that 276,748 houses are vulnerable to SLR of 7 cm by 2030, which will increase to 338,178 houses vulnerable to SLR of 27 cm by 2060, as shown in Table 6.2 [10]. Sea level rise will lead to seawater intrusion which

Table 6.2 Number of housing units vulnerable to SLR

SLR scenario	2030	2060
Low	260,505	273,118
Middle	276,748	338,178
High	281,905	1,110,793

Source: UNDP

Table 6.3 CC impact on Nile Delta according to high SLR scenario

Region	SLR 2030 (cm)	Loss of agricultural lands (km^2)	Loss of agricultural lands (%)	SLR 2060 (cm)	Loss of agricultural lands (km^2)	Loss of agricultural lands (%)
Northeast Delta	27.9	379.3	25.7%	109.6	774.3	52.7%
North Middle Delta	14.8	84.3	1.6%	60.3	523.9	10.4%
West Delta	13	6	0.1%	55	625.6	13.2%
Total of deltas	–	469.6	4.2%	–	1923.8	17.1%

Source: After UNDP

will have negative impacts on water quality [11]. The northern coast of Egypt is threatened by coastal erosion [12], and the Nile Delta is already subsiding at a rate of 3–5 mm per year. Alexandria city will lose about 30% of the city's area due to inundation. Over 1.5 million people will be affected and will lose 195,000 jobs as well as lands and properties in the range of $30 billion if sea level rises to 0.5 m [13]. According to the United Nations Development Programmes, the relative SLR for North Middle Delta in high SLR scenario is estimated to be about 110 cm by 2060 which will cause loss of 52.7% of agricultural lands. In addition, in North Middle Delta, a rise of 60 cm will cause a loss of 10.4% of agricultural lands, while in West Delta, a rise of over 55 cm will cause a loss of 13.2% of agricultural lands. Table 6.3 indicates that the Northeast Delta is the most vulnerable area to sea level rise.

6.2.2.3 Climate Change Impact on Infrastructure

Inundation along coastal areas, due to sea level rise, leads to the damage of property and disruption of infrastructure. An elaborate example is the International Coastal Road, which is considered an important corridor link between the western and eastern borders of Egypt as it runs from Rafah to Matruh (two Egyptian cities), as coastal erosion and sea level rise will eventually damage the aforementioned road [14].

Figure 6.8 shows cross-shore transects along vulnerable low-lying areas along delta coastline. High temperatures and coastal erosion will increase the required maintenance of paved roads as high temperatures cause cracking of road surfaces. This is supported by the fact that during the 1980 heat wave, hundreds of miles of highways were damaged in the United States [15].

Projected floods in Egypt will damage property, harm infrastructure and displace people. A flooding on January 2010 affected over 3500 people and claimed 12 lives [16]. The annual loss in property values due to sea level rise will reach 1–2 EGP billion by 2030 and 7–16 EGP billion by 2060 [10]. Figures 6.9 and 6.10 show a timeline of recent flood events in Egypt.

Moreover, a 50 cm rise in sea level would threaten two million people and force most of them to move away from risk zones in coastal region (Fig. 6.11). Migration waves, due to floods, will go to settle in urban centres and rural settlements, which in turn will increase pressure on cities' infrastructure that is already suffering from many problems due to population growth and low quality of infrastructure.

Displaced individuals who cannot find adequate accommodation will build their own makeshift shelter in slums which will result in an increase of rural settlement problems. In addition, saltwater intrusion will affect foundations of buildings, state of roads and quality of water [17].

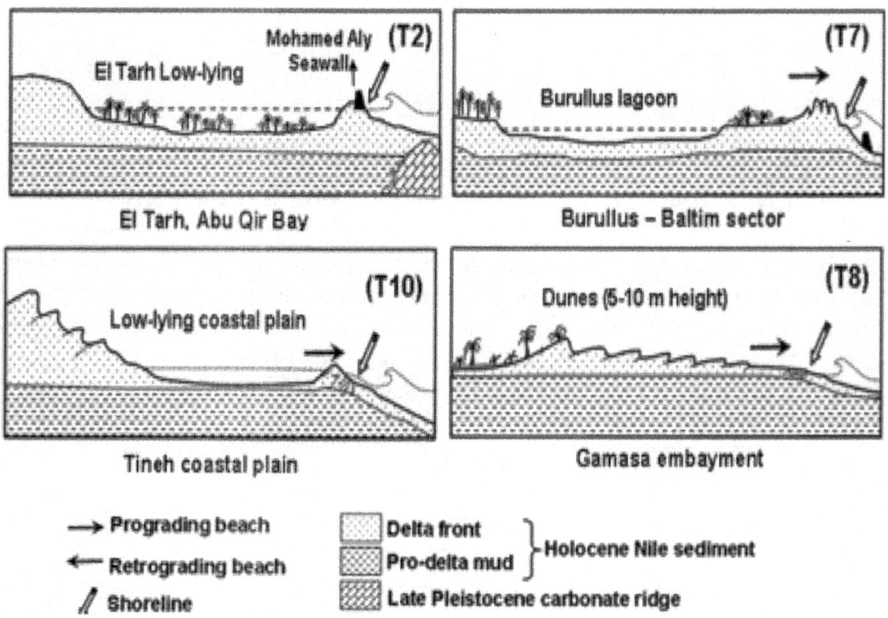

Fig. 6.8 Cross-shore transects along coastline of Alexandria and the Nile Delta. (Source: Omran E. Frihy, Springer)

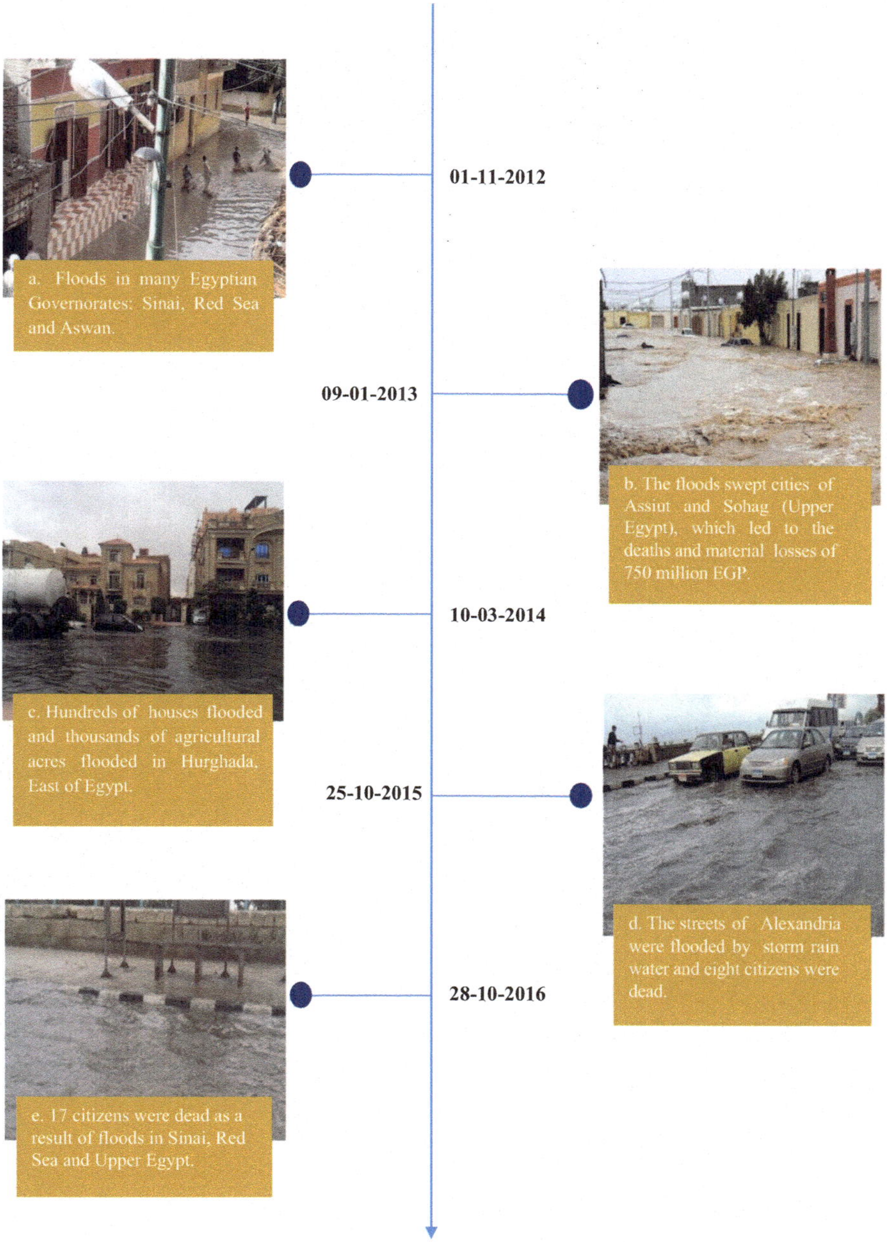

Fig. 6.9 Timeline of recent flood events in Egypt, 2012–2016. (Source: alkessa.com/event.php?id=85 (Accessed on: August 14, 2017))

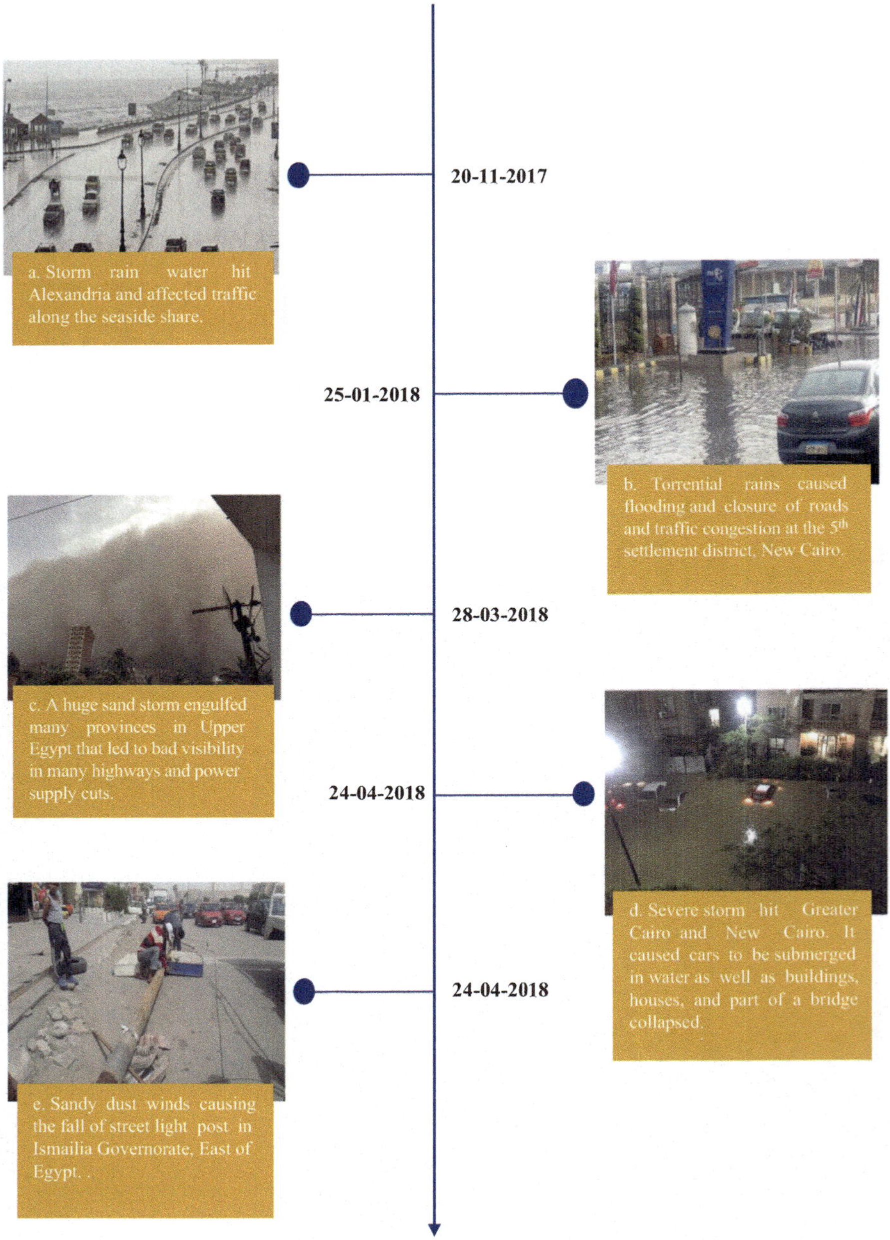

Fig. 6.10 Timeline of recent floods and severe events in Egypt, 2017–2018. (**a**) https://www.egyptindependent.com/heavy-rains-in-alexandra-and-halayeb-on-monday/. (**b**) http://www.egypt-today.com/Artrmicle/1/41036/Wave-of-bad-weather-hits-Egypt. (**c**) http://strangesounds.org/2018/03/sky-turns-orange-after-gigantic-sandstorm-engulfs-southern-egypt.html. (**d**) http://www.egypttoday.com/Article/1/48617/In-pics-Heavy-rains-impact-buildings-throughout-Cairo. (**e**) http://www.egypttoday.com/Article/1/49037/Live-updates-pics-Rain-sand-hit-Egypt-gov-t-hopes\. (Sources: (Accessed on: September 1, 2018))

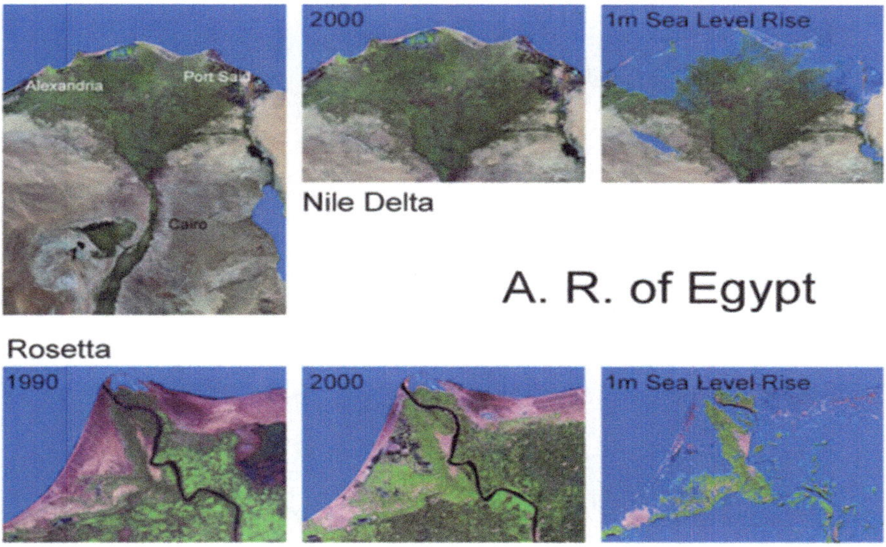

Fig. 6.11 Inundation zones for a 1 m SLR in the Nile Delta and Rosetta. (Source: Remote Sensing Centre, Cairo)

Table 6.4 Projected change in mean annual Nile flow (UNDP)

General circulation model	Egypt allocation 2000 (BCM)	2030 (BCM)	2060 (BCM)
Increased flow	55.5	63.1 (14)	70.6 (27)
Small decreased flow	55.5	52.3 (−6)	49.1 (−12)
Large decreased flow	55.5	45.5 (−18)	35.6 (−36)

Value in parentheses is % change in flow

6.2.2.4 Climate Change Impact on Water

Egypt is vulnerable to water stress, due to climate change, which rapidly increases the gap between water supply and demand. By the year 2020, water demand will have increased by 20% due to the increase in population. The total population number of Egypt grew exponentially by the addition of 36 million from 1950 to 2010, and population growth will continue increasing to reach around 120–150 million by 2050 [18]. Water consumption by agriculture sector represents 85% of the total consumption of the annual total water resource, and large amount of it is wasted due to traditional irrigation systems [19]. Climate change impacts will affect water resources in Egypt, especially the Nile River which supplies 95% of Egypt's total water; optimistic scenarios project a flow increase by one-fourth, and pessimistic scenarios project a flow decrease by more than one-third as shown in Table 6.4 below.

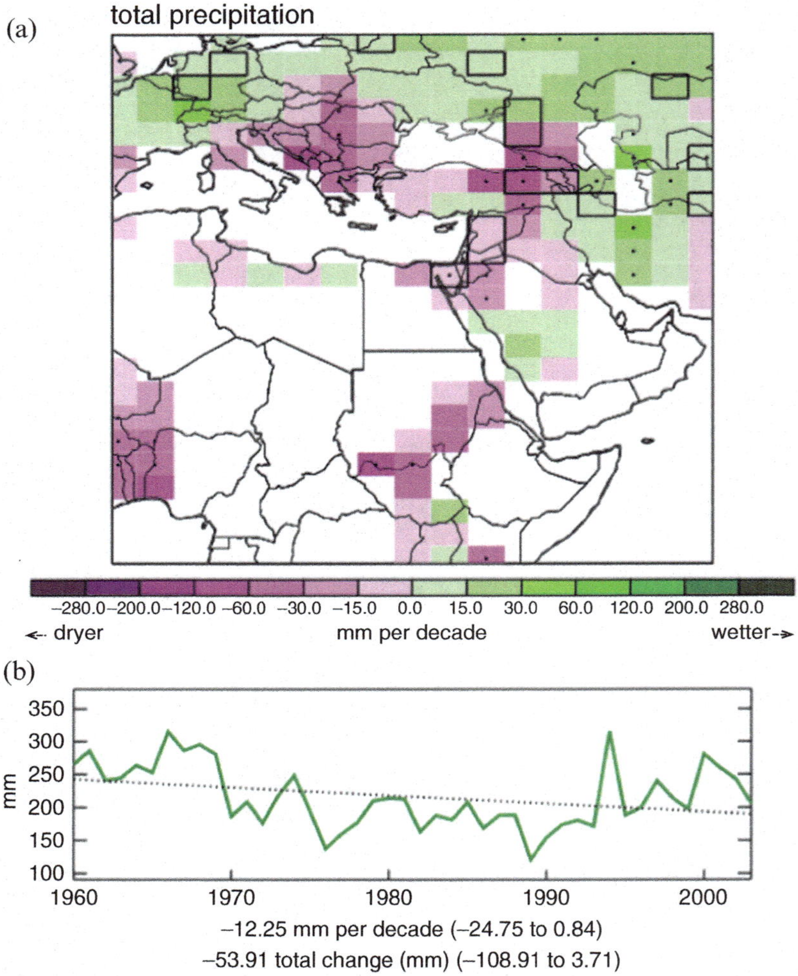

Fig. 6.12 Annual precipitation in Egypt. (**a**) Change in annual precipitation. (**b**) Total annual precipitation. (Source: the UK Met Office)

Furthermore, 70% of studies predicted a decline in Nile water availability [20]. Climate change will decrease the share of water from 700 m^3 to 350 m^3 in annual per capita by 2040, reduce the rainfall on Egypt's Mediterranean coast by 50% (Fig. 6.12a) and decrease the total annual precipitation in Egypt by 54 mm over the period 1960–2003 (Fig. 6.12b). Seawater intrusion increases water salinity [21]. In 2014, already 1% of Egypt's population had no direct access to safe drinking water [22], and climate change will additionally increase the number of slum areas and informal settlements without access to safe water and sanitation.

Table 6.5 Projected changes in crop production in Egypt

	Change %	
Crop	2050	2100
Wheat	−15	−36
Rice	−11	
Maize	−19	
	−14	−20
Soybeans	−28	
Barley	−20	
Cotton	+17	+31
Potato	−0.9 to −2.3	+0.2 to +2.3

Source: UNDP

6.2.2.5 Climate Change Impact on Agriculture and Green Areas

Agriculture is an important sector of economy, representing 13.7% of gross domestic and employing more Egyptian workers than any other sector does, as it provides 30% of all employment positions. Most of the agricultural productivity comes from agricultural lands in Mediterranean coast and Nile Delta; however, the rise in temperature will change crop production as shown in Table 6.5. Agriculture will suffer from water stress, water demand and increased pests and disease, which will lead to increased pressure on irrigation systems due to negative impact of climate change on water resources. Inundation along coasts and low-lying areas due to sea level rise threats agricultural lands along the Mediterranean coast and Nile Delta, and by 2050 [10] such impact will lead to reduced crop production, loss of agricultural lands and increase in soil salinization by water shortage and sea level rise [23].

6.2.2.6 Climate Change Impact on Health

Egypt already suffers from air and water pollution; diseases that are caused by poor water quality and sanitation kill 17,000 children per year. The concentration level of air pollution in urban centre of Greater Cairo is 82.6 mg/m^3, while the World Health Organization standard is 10 mg/m^3. This difference contributes to 20,000 premature deaths yearly in Cairo and Alexandria and puts pressure on Egypt's economy as it costs from 3% to 6% of GDP.

Climate change could increase air pollution due to high temperature and low precipitation leading to the increase in annual mortality by thousands and to the loss of 10–24 billion EGP per year as shown in Table 6.6 [10]. Air and water pollution levels in Egypt are high and contribute in many premature deaths and morbidity. About 17,000 children per year, one-fifth of all childhood deaths, are estimated to result from poor water quality, inadequate hygiene and poor sanitation. These conditions also contribute to high levels of infectious diseases among children and adults, particularly diseases caused by worms and other parasites that result in a

Table 6.6 Estimated increase in annual mortality in Greater Cairo from a 1 μg/m³ change in PM

Health effect	2060	Economic value (billion EGP)
Optimistic socioeconomic scenario		
Low $PM_{2.5}$ estimate (adults)	708	10.7
High $PM^{2.5}$ estimate (adults)	1610	24.2
Pessimistic socioeconomic scenario		
Low $PM_{2.5}$ estimate (adults)	1015	6.3
High $PM_{2.5}$ estimate (adults)	2308	14.2

Source: UNDP

Table 6.7 Estimated annual mortality in Greater Cairo from heat stress

	2030			2060		
	CGCM63	ECHAM	MIROC-M	CGCM63	ECHAM	MIROC-M
Optimistic	662	662	736	1662	1579	1924
Pessimistic	722	722	802	2302	2187	2665

Source: UNDP

yearly loss of 615,000 disability-adjusted life years, i.e. rate of deducted years from a person's life [24].

Global warming increases heat stress, particularly in urban areas such as Cairo and Alexandria, which leads to illness and thousands of deaths due to heat stroke as shown in Table 6.7. Higher temperatures, water stress and malnutrition increase bilharzia, malaria, rift valley fever, tuberculosis, avian influenza and diarrhoea. Moreover, respiratory patients will increase due to the decrease in air quality by climate change. A decrease in crop production due to climate change will increase malnutrition, and the spread of these diseases will lead to lower productivity and economic losses [25].

6.2.2.7 Climate Change Impact on Ecosystem

The Mediterranean coastal shoreline of Egypt contains five large lakes making up about 25% of the total wetlands in the Mediterranean region. Low-lying coastal zones and Nile Delta are exposed to rising sea levels. The land subsidies in coastal zones are at least 1.6 millimetres per year at Alexandria, 1.0 mm per year at Burullus and 2.3 millimetre per year at Port Said which would exacerbate the impacts of rising seas and maybe exceed the expected value of 18–59 cm by the end of this century. Seawater intrusion and increasing soil evaporation due to high temperature play a main role in the salinity of almost 35% of the agricultural lands in Egypt which harm ecological system [26].

Erosion due to SLR is affecting Damietta city's coastal zone which led to retreating more than 500 m between 1983 and 1995. Also, Aswan Dam plays a role in reducing sediment in Nile Delta and increasing vulnerability to coastal erosion [27], but the biggest danger that accelerates the pace of climate change impact is anthropogenic pressures. In Lake Manzala, human-induced acts such as conversion of wetland ecosystems to agricultural lands, land use change, lack of appropriate institutional management systems and the discharging of considerable amounts of sewage and industrial wastes directly into the wetlands and lakes increase vulnerability of ecosystem. An increase in sea level rise over time is expected to increase the rate of saline seepage of the topsoil of the delta and may have a serious impact on agriculture and groundwater resources in the upper Nile Delta. Moreover, an increase in the salinity of Lake Manzala may lead to negative impacts on ecology and fisheries [28].

A substantial number of the currently endangered species, shown in Fig. 6.13, might be lost resulting from climate change impacts in Egyptian northern lake habitats such as the lagoons of Maryut, Edku, Burullus, Manzala and Bardaweel [29]. The northern part of the Nile Delta Natural Protectorates of the Bays of El Gamil, Ashtoon El Gamil and Al Manzala lagoon in Port Said Governorate represents wetlands and natural restricted area for birds that might be affected by sea level rise. Similarly, in El-Brolus protected area in Kafr El Sheikh Governorate, a habitat of almost 135 amphibious plant species has a significant role in receiving immigrating wild birds and is also exposed to the negative effects of SLR. Also, in Zaraniq Protected Area and El Bardwaeel Marsh in the North Sinai Governorate are key gathering points for birds' migration in the world, which are considered habitats for living and reproduction; over 270 species of birds have been recorded in the area such as pelicans, herons, storks, crested lark, quail, white stork, lark, avocet, Hem Harrier Rail and falcon [30].

Coral reefs in the Red Sea have undergone significant decline due to coral bleaching as a result of a change in water temperature and acidity, in addition to human-induced pressure. They are damaged and polluted due to the high volume of

Fig. 6.13 Vulnerable biodiversity in northern lakes, Egypt. (Source: Egyptian Environmental Affairs Agency)

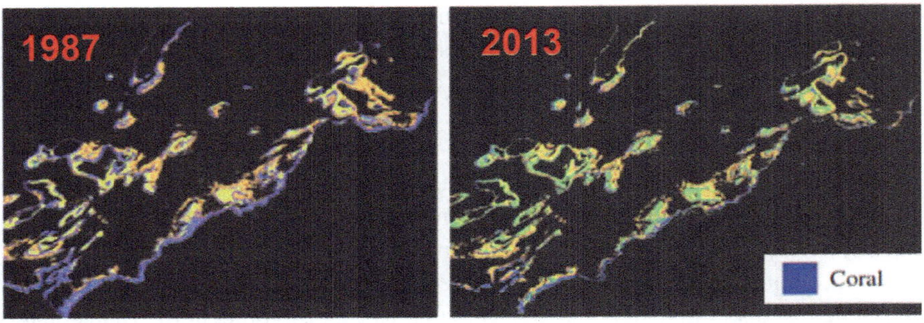

Fig. 6.14 Coral bleaching between 1987 and 2013. (Source: Mohamed El Raey, MARCH 2014)

commercial shipping, oil spills, sewage and nutrient loading from hotels and resorts and overfishing [31]. In Hurghada, Egypt, coral reefs had a declining rate of 40% during 1987–2000 (Fig. 6.14) and also face more drastic declines of 49% and 45%, respectively, between 2000 and 2013, which affects biodiversity in the Red Sea that already contains over 1000 species of mangrove vegetation in the Red Sea coasts [32].

6.2.2.8 Climate Change Impact on Fisheries

Egypt produces 93% of its fish needs from Nile River, the Mediterranean and Red Seas, the northern lakes as well as fisheries in the north of the Delta. Increasing temperature and salinity of sea will have an effect on coastal lake fisheries as it forces fish to migrate [33]. Furthermore, water stress will affect aquaculture productivity in Egypt which represents 80% of fish production. The value of fisheries from coral reef was 220 million EGP in 2000, and an 80% reduction in coral reefs due to climate change will cause economic losses estimated to be 165 million EGP [10].

6.2.2.9 Climate Change Impact on Tourism

Tourism accounted for 11% of GDP and 14.4% of foreign currency and employed about 12% of Egypt's workforce in 2010 [34]. It is estimated that 84.5% of the visits to Egypt are for leisure and recreation. Recreational expenditures on the Red Sea coral were US$472 million (2.6 billion EGP) in 2000, and 75% of tourism activity in Egypt is concentrated on Sinai and the Red Sea. The rise in temperatures and increased seawater acidification increase coral reef bleaching (Fig. 6.15), as it is expected for coral reef bleaching in the red sea to increase by 80% by 2060. It is important to note that coral reef growth has decreased by 30% in 2013 [10]. Loss in internal beach tourism will occur due to sea level rise in Alexandria, Port Said, Ras El Bar, Gamasa and Baltiem by 2025 [33]. Sea level rise threats many areas such as archaeological sites in Abu Keer, Qaitbey Citadel and Selsela [35]. Figure 6.16 shows some of the archaeological sites in the eastern harbour of Alexandria. In

Fig. 6.15 Healthy (left) and bleached (right) coral reef. (Source: NOAA)

Fig. 6.16 Archaeological sites in eastern harbour, Alexandria. (Source: Al-Ahram Weekly)

addition, heat waves will affect attractiveness and tourism movement in Egypt because most of the tourists prefer moderate weather [36].

6.2.2.10 Climate Change Impact on Politics

Egypt suffers from water poverty as the per capita share of water decreased from 2526 cubic metres in 1947 to 663 cubic metres in 2014; also, water demand increased by 23.7% during the period 2002–2012 [37] and expected to fall below 582 cubic metres by 2025 due to climate change [38].

The decrease in water resources of Egypt increases water stress and could increases tension among Egypt and the Nile Basin countries, such as Ethiopia, which led to political concerns due to the construction of the Grand Ethiopian Renaissance Dam that reduces water quotas of Egypt. Sudan can also be mentioned as an important part of the dispute over water quotas between the Nile Basin countries. However, actions have been taken in Egypt to conserve water use and raise awareness of citizen to reduce water waste.

Table 6.8 (below) summarizes climate change risks on Egypt that mainly counts 32 risks distributed among sectors. One-third of total risks already occurred, and the rest will occur between 2030 and 2100 according to CC scenarios, as shown in Table 6.9.

Table 6.8 (a) Climate change risks on Egypt by sector (1–2)

Sector	Risk
Water resources	Decline in annual per capita share of water from 700 m^3 to 350 m^3 by 2040 projected [39]
	Changes in the flow of the Nile [33] [20]
	Possibility of a 50% reduction of rainfall on Egypt's Mediterranean coast [21]
	Increase gap between water supply and demand [39] [38]
	Groundwater could experience increased levels and salinity due to sea level rise and sea water intrusion [21]
Agriculture	Reduced crop production [21] [39] [16]
	Loss of agricultural land to desertification [39] [40]
	Increased pests and disease [33] [10]
	Increased pressure on irrigation systems [10]
	Increasing soil salinity [39] [41] [23]
	Increased livestock stress, as well as stress on fodder crops [21]
Coastal zones	Coastal erosion [12]
	Inundation along Mediterranean coast and Nile Delta [10]
	Extreme events of heat waves, sand and dust storms [21] [33]
Tourism	Sea level rise threat archaeological sites in Abu Keer, Qaitbey Citadel, Selsela (eastern harbour) in Alexandria and the Rosetta [33]
	Heat waves will effect on attractiveness and tourism movement in Egypt [39]· [41]
	Loss in beach tourism according to sea level rise in Alexandria, Port Said, Ras el Barr, Gamasa and Baltiem [39]· [41]
	Loss in beach tourism due to Coral reefs bleaching in the Red Sea [33], [41]· [10]
Urban Areas	Damage of property and disruption of infrastructure and displace people due to sea level rise [21]
	Increase heat island effect and decrease air quality in urban cities [21]
	High temperatures and coastal erosion will increase maintenance of paved roads [33], [15]

Dark Red font represents risks already occurred.
Black font represents risks that will occur in the future
[ ] The red number between two red bracts is reference

Table 6.8 (b) Climate change risks on Egypt by sector (2–2)

Sector	Risk
	Damage roads like the international coastal road from Rafah to Matruh due to sea level rise [33], [42]
	Migration waves due to floods will go to urban centers and rural settlement, will lead to increase pressure on cities infrastructure [43]
	Saltwater intrusion will affect foundations of buildings, roads and water quality [21]
Health	Higher temperatures, water stress and malnutrition increase bilharzia, malaria, rift valley fever, tuberculosis, avian influenza and diarrhea [21] [33]
	Respiratory patients will increase due to decrease Air quality by Climate change [10]
	Increase malnutrition due to decrease in crop production [21]
	Heat waves that cause non-comfort conditions, illness and deaths related to heat stroke particularly in urban areas [33]
	Spread diseases will lead to lower productivity and economic losses [21]
Fisheries	Coral reefs bleaching will decrease fisheries productivity in red sea [10]
	Water stress and water salinity will affect Aquaculture productivity [10] [33] [39]
Political	Decrease of water resources increases friction among countries sharing the same water resources (e.g. Nile and Euphrates) and leads to political unrest [39] [38]

Dark Red font represents risks already occurred.
Black font represents risks that will occur in the future
[ ] The red number between two red bracts is reference

Table 6.9 Classification of CC risks on Egypt

Occurrence	Risks' number
Risks already occurred	9
Risks that will occur in the future	23
Total of risks	32

Table 6.10 Ranking of CC vulnerabilities in Egypt

Resources/risk ranking	Risk	Certainty of impact	Severity of impact	Importance of resources
Coastal resources	SLR Coastal erosion Soil salinization	High to medium	High	High
Water resources	Decrease Nile flow Low precipitation Saltwater intrusion	Medium	High	High
Agriculture	SLR Soil salinization High temperature	Medium to low	Medium to low	High to medium
Energy resources	High temperature Decrease Nile flow	Medium to low	Medium to low	Medium to low

Source: OECD

As noticed, a huge number of CC risks affect all of the sectors and activities in Egypt, as it is one of the most vulnerable countries to climate change, especially due to sea level rise that threatens over two million of the coastal population. Also, CC risks slow rate of development.

Therefore, Egypt provides a rich reference case for developing countries in this chapter.

Table 6.10 lists the ranking of CC vulnerabilities in Egypt. The potential impacts of climate change on coastal resources are ranked as the most important due to sea level rise and coastal erosion. High-rate land subsidence due to reduced sediment by Aswan High Dam increases the vulnerability to inundation. Also, high coastal sensitivity due to high population, density and centralized development project on waterfronts increases lives and economic losses. Second important are the risks on water resources as it provides 97% of Egypt's water demand. Risks such as decrease in Nile flow, low precipitation and saltwater intrusion will in turn decline the share of water in terms of annual per capita. Third in importance are risks on agriculture as this sector employs over 28% of the labour force according to the World Bank. Sea level rise (SLR) led to the loss of hundreds of agriculture lands, and high temperature will decrease crop productivity which will force a lot of farmers to migrate to urban areas to seek a new life. Lastly, energy resources will drain due to the boost in energy demand due to high temperature and a decrease in hydroelectricity generation, which provides 11% of total electricity generation according to the Electricity Utility and Consumer Protection Regulatory Agency, due to decrease in Nile flow. All of these risks make Egypt one of the most vulnerable countries to CC, and as a result, the next section investigates policies of climate change adaptation in developing countries taking into consideration Egypt as a reference case to tackle the aforementioned risks.

Box 6.2 Climate Change Risks in Egypt

1. Urban and coastal areas
   - Coastal areas are the most vulnerable areas in Egypt due to sea level rise which mean urban and economic loss and migration waves due to inundation.
2. Water sector
   - Decline in the share of water due to changes in the Nile flow and seawater intrusion.
3. Agriculture sector
   - Reduced crop production due to desertification, soil salinity and high temperature.
4. Tourism
   - SLR threats archaeological sites in eastern harbour in Alexandria.
5. Health
   - In urban areas, heat waves that cause illness and deaths and decreased air quality will threat respiratory patients particularly.
6. Political
   - Political conflict will arise between Nile Basin countries on water.

6.3 Climate Change Adaptation Policies in Developing Countries

Policies for climate change adaptation in developing countries are vital to assist governments and private sectors to cope with climate change risk amid the huge challenges and limited resource in these countries especially after 2009. The National Adaptation Programmes of Action (NAPA) was created during the Conference of the Parties (COP) 7 in Marrakech, Morocco, in 2001. The programme prioritizes climate change adaptation activities addressing the urgent and immediate needs and concerns of the least developed countries (LDCs) [44]. Till 2018, 51 of the LDCs have submitted their adaptation plans to UNFCCC [45]. The plans draw climate change adaptation policies.

6.3.1 *Classification of the Climate Adaptation Policies*

6.3.1.1 Settlement and Infrastructure Policies

Municipal and local governments have to consider CCA through preparing and updating regularly the Local Climate Change Action Plans (LCCAP). In the Philippines, for example, local government units streamlined CCA policy by climate action planning. As most of the built environments existed before, CCA policies and urban polices pay attention to retrofitting of existing buildings. In Brazil, the City Council of Recife adopted Municipal Green Roof Law that enforces the roofs of all buildings of more than four floors as well as commercial buildings of more than 400 square metres be covered in native vegetation. This initiative promotes rainwater capture, energy efficiency, sustainable design, and the use of recyclable materials in buildings [46].

Recently, urban policies directed to be 'transformative' by considering social justice and addressing exposure to CC risks. National Urban Development Policy (NUDP) in Nigeria empowers disadvantaged groups through households to begin treating land as an asset and use it to build wealth through win-win approach [47]. The National Urban Policy addresses CCA by decreasing the sensitivity and increasing the adaptation capacity. For example, Uganda improved its administrative and legal environment for urban planning to correlate between urban development and urban poverty alleviation [48].

In Asia, the city of Hat Yai in Thailand adopted a participatory approach to adapt to severe floods. The community-based flood preparedness in the city involves communities in all stages of adaptation to build self-reliant and long-term climate change adaptation. The project strengthens community-based preparation for floods in the assessment stage. Also, local communities provided information to fill the information gap, and authorities raise their awareness about climate implications and flood-related risks. In the planning stage, local communities develop their own flood plans in consultation with authorities, which promotes local ownership and community's self-reliance [49].

In addition, many of the developing countries established their national climate action plans (NCAPs). However, on the city scale, the climate action plans are in early stage. For example, in Asia only 29 of the 894 cities have climate change plans (CCPs). Also, 36% of 29 cities' plans focused on adaptation and mitigation, and 36% concentrate on mitigation only whereas 28% on adaptation only. The analysis of CCPs by sectors relevant to infrastructure investments found out that 79% include energy supply and distribution followed by buildings and water services (59%), planning and urban land use (48%), transport (48%), flood protection (41%), outdoor lighting (31%) and waste management (24%) [50].

6.3.1.2 Water Resources Policies

Water management plays a vital role in climate change adaptation concerning the water sector. In Phnom Penh, Cambodia, instituting efficient management practices at the Phnom Penh Water Supply Authority (PPWSA), through good governance has led to an increase of the daily production of drinkable water by 360%. Reduced corruption has also helped the PPSWA to increase revenues and significantly provided access to clean water to more residents of Phnom Penh. Urban governance and infrastructure improvement in Khulna, Bangladesh, support capacity development for alternative water and sanitation plans, improve service delivery of safe drinking water for the city's informal settlement and support a sustainable pro-poor transportation plan as the infrastructure is the interlinked set of physical buildings and engineered systems that support other sectors or function within urban areas [51].

6.3.1.3 Ecosystem and Tourism

Usually ecosystem interlinked with tourism in Small Island Developing States. Adaptation pilot projects in Maldives and Fiji focus on integration of tourism into national adaptation strategies; the projects deal with tourism in an integrated way, recognizing their strong interconnections and mutual interests where tourism can generate benefits through it. The projects form part of national strategies for both climate change adaptation and sustainable tourism development, supporting a number of related sectors and policy areas, such as ecosystem. Pilot projects improved inter-ministerial coordination, streamlined regulations and environmental impact assessment processes, improvement of climate information for long-term strategic tourism planning as well as seasonal and short-term operations and activities [52].

6.3.1.4 Agricultural and Green Spaces Policies

A lot of developing countries depend in agriculture as one of the most important income sources; therefore, the climate action plans address agriculture to support African agriculture in the face of climate change incorporating a multidisciplinary set of stakeholders including climate science experts, agricultural practitioners and technicians, local communities/civil society, donors and policy-makers [53]. In South Asia, adaptation policy focuses on adaptive capacity in the context of agriculture systems by engaging multiple stakeholders and integrating initiatives with government priorities to generate politically to be sustainable and institutionalized [54].

6.3.1.5 Health Policies

Health policies reduce negative impacts of climate change on health. These policies assist in building national capacities to analyse climate-induced risks to health and to identify adaptive, preventive, and curative measures. Also, such policies support the review process of operational aspects, such as institutional structures, capacities, and financial and budgetary planning processes [55]. In Uzbekistan, policies increase adaptation capacity of health-care system in these provinces to cope with climate-induced diseases by equipping health-care personnel and the wider population with essential tools and knowledge to prevent the detrimental effects of climate on human health [56].

6.3.2 Climate Change Adaptation Policies, Egypt: Case Study

6.3.2.1 Settlement and Infrastructure Policies

Egypt adaptation policies on coastal areas aim at reducing climate change-associated risks, especially sea level rise by some tools such as mapping vulnerability of coastal area to sea level rise and monitoring sea level rise, as well as the promotion of technical and scientific capacity to assess threats, losses and cost-benefit analysis. Moreover, rehabilitating infrastructure for adaptation to sea level rise by expanding protected zone through the utilization of Mohammed Ali seawall to include unprotected zones (Fig. 6.17) is another adaptation policy approach. Other approaches include (1) using other engineering and soft measures, (2) relocating or retrofitting vulnerable buildings and infrastructure, (3) increasing resilience of coastal areas by depending on sustainable development and (4) enhancing social-economic condition of poor people.

The government aims to amalgamate the efforts of all institutions to work within national adaptation plans and build up the Egyptian community's capability to adapt [33]. The adaptation methods include assessing the capabilities of community and developing programmes to be flexible to future risks through forming partnership with donors for developing a participatory adaptation strategy for informal areas, raising awareness among local residents and civil society and depending on small-scale development measures in informal areas using participatory approach of local residents [57] to upgrade informal areas and strengthen adaptive capacity. This determines clear priorities for the implementation of adaptation programmes through raising awareness.

Figure 6.18 depicts heat wave mapping that illustrates the urban agriculture and green open spaces depending on renewable energy, green home retrofitting and green micro-insurance for offsetting climate change risks [58].

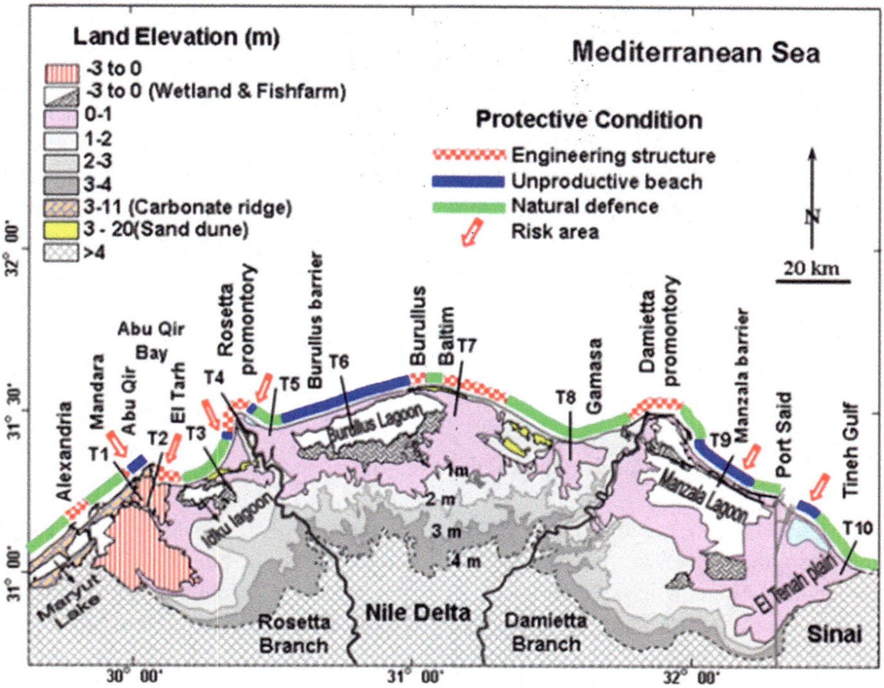

Fig. 6.17 The lower coastal plain of the Nile Delta and Alexandria. (Source: Omran E. Frihy & Mahmoud Kh. El-Sayed, Springer)

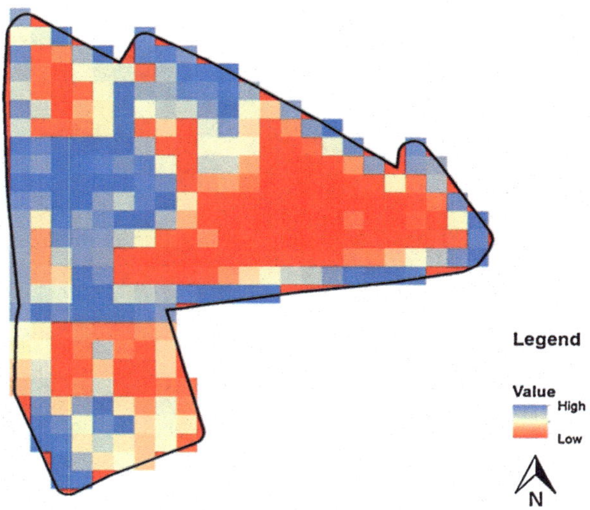

Fig. 6.18 Mapping heat release potential Ezbet El-Nasr, Cairo. (Source: GIZ)

6.3.2.2 Water Resources Policies

The Egyptian Ministry of Water Resources and Irrigation (MWRI) has public awareness programmes aimed to inform the public, through different channels, to raise awareness on climate change and water-saving techniques, achieving public participation in water policies and programmes [59]. Egypt set up Inter-Ministerial Committee on Climate Change to consider adaptation into national planning, build up models to simulate climate change impact for build-up adaptation scenarios, manage Aswan Dam and reservoirs to adapt to possible changes in water runoff and support researches and studies that deal scientifically with uncertainty [60]. Figure 6.19 shows modern irrigation method used in Toshka Valley, Cairo, but until now, most of agricultural lands depend on traditional irrigation methods.

6.3.2.3 Ecosystem and Tourism Policies

The new policy includes techniques such as expanding the protected marine areas and strengthening environment preservation regulations, developing coastal zones and encouraging sustainable and eco-tourism, evaluating the vulnerability of tourism sites to reduce climate change risks in touristic areas, redirecting tourism activities to less vulnerable sites and ensuring effective monitoring of implemented adaptation plans. Also, the policy has a significant role in stakeholders' implementation of adaptation by raising environmental awareness of local communities [61].

Fig. 6.19 Modern irrigation method in Toshka project, New Valley. (Source: GIZ)

6.3.2.4 Agricultural and Green Spaces Policies

Regarding agriculture and green spaces policies, some new actions have taken place. These policies include building institutional framework for managing crises and disasters resulting from CC affecting agriculture. Such policies would consequently assist in assessing and monitoring the threats and deriving recommendations as well as increasing the efficiency of irrigation and using new technologies. Also, these policies help in enhancing crop productivity, modernizing and diversifying agricultural production to be able to realize maximum productivity. In addition, neutralizing the change in climate indicators for food security improvement, reviewing policies of land use and assessing vulnerability of agricultural land in the delta resulting from human-induced pressure and sea level rise-induced pressure are vital. Moreover, improving land management and enhancing adaptive capacity of small farmers are necessary policies and actions in order to strengthen their capabilities to adapt to climate change impacts on land use, plants, and animal production.

Furthermore, monitoring and assessing the vulnerability of animal production, developing vaccine to combat expected diseases and improving livestock production are all methods used in animal and livestock criteria. As for water species, some of the methods being used are protecting fish from invasive species that disturb environmental balance of the fish wealth in Egypt, protecting fish wealth from infections and diseases, improving the aquaculture ecosystem, achieving diversity of animal and fish to achieve food security for the community and developing new system of aquaculture and animal production flexible enough to be implemented under CC [33]. Figure 6.20 shows urban agriculture on rooftop of an informal building in Cairo (Fig. 6.20a) and a green roof of a farmer prototype house in Beheira Governorate (Fig. 6.20b).

6.3.2.5 Health Policies

Egypt's adaptation plan in health sector includes reduction of socio-economic vulnerabilities, enhancement of national public health infrastructure, improvement of health services quality and the expansion of accessibility to quality health services,

(a) (b)

Fig. 6.20 Examples of top roof green farming in Egypt. (**a**) Green roof – an informal building, Cairo. (**b**) Green roof – a farmer building, Al Karz village, Beheira Governorate. (Image Source: **a**. pinterest.com (Accessed: April 4, 2016), **b**. Lead author)

Table 6.11 Climate change adaptation policies in Egypt by sector (UNDP) [33]

Sector	Policy
Settlement and infrastructure	Relocating or retrofitting vulnerable buildings and infrastructure
	Depending on sustainable development and enhancing social-economic condition of poor people
	Enhancing national partnership in CC adaptation and building up technical and scientific capacity
	Developing a participatory adaptation strategy and depending on small-scale development measures
	Mapping vulnerability of coastal area to sea level rise and monitoring sea level rise
	Expanding protected zone by Mohammed Ali seawall
Water resources	Raising community awareness on climate change and water saving
	Considering adaptation to climate change and disaster risk reduction, a primary national priority
	Building up models to simulate climate change impact to adapt to possible changes in water runoff
Agriculture Sector	Depending on varieties capable of achieving maximum productivity and neutralizing change in climate indicators for ensuring food security
	Development of new flexible agro-economic systems and structures to manage crops, fish farming and animal production
	Increasing the efficiency of irrigation and protecting land from degradation
	Reviewing policies of land use and improving land management
	Enhancing adaptive capacity of small farmers to deal with change in land use and internal migration
Health sector	Improving vaccination programmes and control programmes for infectious diseases to adapt to CCR on health
	Raising community awareness about climate change risks and means of adaptation
	Increasing the efficiency of the health care and enhancing air quality
Ecosystem and tourism	Mapping vulnerability and redirecting tourism activities to less vulnerable sites
	Incorporating disaster risks within Egypt's sustained tourism development plans
	Expanding the protected marine areas, strengthening environment preservation regulations and encouraging eco-tourism

development of research institutes to deal with epidemics resulting from CC and improvement of vaccination programmes and control programmes for infectious diseases. Other policies include developing early warning systems and raising community awareness about CC threats and adaptation [62]. The United Nations Development Programmes in partnership with the Egyptian Environmental Affairs Agency launched 'Low Emission Capacity Building project' in 2013 to design a low-emission development strategy for ensuring environmental sustainability and enhancing air quality [63].

Table 6.11 investigates the coverage of climate change adaptation policies among sectors such as settlement and infrastructure, water resources, agriculture, health, ecosystem and tourism to underline any shortcomings in policies. The data relies on the official reports of the Ministry of Environment in Egypt and reports of funders and donors of adaptation projects such as the World Bank.

These policies are adequate to the adaptation in different sectors and achieve tangible results. Egypt is the first country in MENA region to sign environmental agreements such as the Framework Convention on Climate Change and the Kyoto Protocol. Since 1997, Egypt had a number of environmental and sectorial plans and committees that are responsible to CC, but the actual implementation of the CC adaptation plans are facing several obstacles. These barriers are classified as financial, social and cultural, information, technological, and development, as shown in Table 5.4 of Chap. 5.

Nevertheless, the effectiveness of the policy implementation on the ground is still weak along with policy conflicts which hinder CC adaptation. The next section identifies climate change adaptation measures in developing countries, and Egypt is selected as a reference case – due to data availability – to demonstrate the adaptation measures taken in the following sectors: coastal areas, urban areas and infrastructure as well as activities in the city such as fresh water, health, agriculture, forest, fishing and biodiversity.

Box 6.3 Climate Change Adaptation Policies in Egypt

1. Urban and coastal areas
   - Coastal mapping of the vulnerability of coastal area to relocating or retrofitting vulnerable buildings and building up capacity by depending on small-scale development measures
2. Water
   - Considering CC adaptation in water management and raising community awareness about water-saving techniques
3. Agriculture
   - Increasing the efficiency of irrigation, improving land management and enhancing adaptive capacity of small farmers
4. Ecosystem and tourism
   - Expanding the protected marine areas, strengthening environment preservation regulations and encouraging eco-tourism
5. Health
   - Increasing the efficiency of the health care and enhancing air quality

6.4 Climate Change Adaptation Measures in Developing Countries

Measures adopted in developing countries for adaptation to climate change severe events and risks are highlighted in this section. These measures cover urban settlements that focus on energy efficiency, water resources and ecosystems, as well as green open space agriculture and public health.

6.4.1 Classification of the Climate Change Adaptation Measures

6.4.1.1 Settlement Measures

A lot of climate change adaptation measures in settlement focus on energy efficiency as one of the most effective and feasible visions for climate change adaptation and mitigation. Puebla City in Mexico became a pilot city for the World Bank for the rapid assessment of city energy. This work jumpstarted measures focused on street lighting, municipal buildings, transit systems and solid waste management in city and contributed to the development of a national urban energy efficiency strategy. In 2015 the city was chosen as Mexico's national winner of WWF's city challenge, in particular for its ambitious commitments and high level of investments in climate action [64].

Coimbatore is one of the 36 cities that finalized its solar city master plans as part of its participation in India's National Solar Cities Programme. The plan reduces conventional energy consumption by 10% within 5 years, through investments in renewables and energy-efficiency measures. The city has installed solar photovoltaic panels (PVs) on all municipal buildings and is completing its energy-saving project for streetlights. The city of Coimbatore, the second largest city in the southern state of Tamil Nadu, has a population of 2.2 million in the greater metropolitan area. The solar city master plan includes comprehensive investments in an array of renewables and energy-efficiency measures both in municipal operations and in the wider community [64].

The informal settlement in Da Nang city, Vietnam, is highly vulnerable to storms and floods because their homes are often poorly constructed and maintained, and families frequently struggle to recover from storm damage due to low income and rising costs. In order to increase the city resilience, the Women's Union in Da Nang administers a revolving loan to make their homes resistant to storms. The building of storm-resistant houses initiative has helped hundreds of lower-income households rebuild and repair their houses. The Women's Union in Da Nang administers a revolving loan to make the urban poor's homes resistant to storms in eight disaster-prone wards. Three hundred twenty households have reinforced or reconstructed their homes in the first 3 years, and up to 420 households have made their homes resistant to typhoons and floods by 2017. Currently, the informal settlements in eight vulnerable wards in Da Nang city are more resilient to the storms that are increasing in frequency and in intensity due to climate change [65]. The adaptation to coastal erosion in Rufisque, Saly and Joal in Senegal against further erosion caused by sea level rise and storm surges aims to reduce exposure to climate change impacts on the coast by protecting houses and coastal infrastructure – such as fishing docks, fish processing plants and even tourism – that are threatened by erosion and saltwater intrusion. Some measures the project introduced include anti-salt dikes to mitigate the calcination of lands and sea defences to prevent coastal erosion. In Rufisque, the construction of a seawall along the coastline will protect historical heritage houses that are being threatened by coastal erosion; the preservation of tourist infrastructures and economic activities in Saly will help maintain at least 3000 direct jobs and 9000 indirect [66].

6.4.1.2 Water Resources Measures

Direct influence of sea level rise and backwater curves has a significant effect on water levels in the Red River which flows through the Hanoi city centre in Vietnam. Climate change projections anticipate an increase in the frequency and intensity of typhoons and tropical storms, which increase the flood risk for urban areas. Integrated river basin management for climate resilience project was implemented in order to reduce Hanoi's vulnerability to climate change. The project takes an integrated approach to enhance the Red River basin's water management. The climate resilience programme entails hard and soft adaptation initiatives to ensure the system's resilience. These initiatives include a revision of current flood protection infrastructure, the creation of additional water retention capacity in the form of reservoirs, enhancement of scientific capacity to respond to emergencies and finally new flood discharge regimes that incorporate socio-economic considerations [67].

6.4.1.3 Ecosystem Measures

Green Coast is a community-based coastal ecosystem restoration programme implemented in Indonesia, Sri Lanka, Thailand, India and Malaysia that aims to increase resilience of coastal areas. The project aims to rehabilitate coastal ecosystem, building sustainable livelihoods, and develop village regulations to support environmental conservation efforts and environmental education. In the areas of Aceh and Nias in Indonesia, it successfully rehabilitated 893 ha of coastal land through the planting of mangroves and beach plants along with the conservation of coral reefs. Restoring coastal ecosystems helps to protect communities and livelihoods from the impacts of climate change [68] as mangroves dissipate the energy and reduce the size of wave heights 5–7.5 times more than non-vegetated beach surfaces [69].

6.4.1.4 Agricultural and Green Spaces Measures

Climate adaptation for rural livelihoods and agriculture in Malawi included enhanced afforestation, conservation of agriculture, drought-tolerant crop introduction, water distribution, irrigation efficiency, groundwater capture, water recycling and water system rehabilitation. These adaptation activities were also designed to help achieve broader development objectives associated with rural poverty reduction, by developing and enabling environment for adaptation; the project aimed to foster replication of these interventions beyond its direct project activities [69].

6.4.1.5 Public Health Measures

Cape Town, South Africa, depends on urban design to enhance the public health in the city including improving surface cover to increase reflectivity, increasing urban tree canopy and generating air movement through urban corridors. Also, Cape Town is considering a comprehensive 'heat-health' action plan including monitoring, emergency medical services and public and professional awareness [70].

6.4.2 Climate Change Adaptation Measures, Egypt: Case Study

6.4.2.1 Settlement Measures

Egypt's primary energy consumption comes from mainly non-renewable resources (about 88%), which accounted for 91% of the country's total energy consumption in 2013; the other 4% were distributed among hydroelectricity energy (3%) that comes from Aswan High Dam and solar and wind energy (2%), which is expected to produce 20% of total power generation through renewable energy in 2020 [71] and 40% by 2035. More renewable energy sources to be added to the grid due to the world's largest solar power plant in Benban in Aswan with the capacity of 2.00 GW when competed – at the moment 200 MW has been installed and officially opened [72]. Egypt allocated interest-free loan to convert tourist resorts on the Red Sea to renewable energy and targeting to switch 100,000 hotel rooms from conventional heating to solar water heating [73]. Also, private NGOs play a role in providing electricity to rural areas such as Tanweer El Heiz Project that provides electricity for 68 homes in El Heiz village near Bahariya Oasis by installing solar PV cell units as shown in Fig. 6.21 [74].

Egypt started reusing wastewater for irrigation since 1930 in sandy soil areas near Cairo, as about 5.5 BCM of drainage water are being reused after mixing with fresh water and this number is expected to increase up to 9.6 BCM in 2017. Treated wastewater is being used in irrigation in an amount of 0.7 BCM/year, while the amount of desalinated seawater in Egypt is estimated to be around 0.03 BCM/year

Fig. 6.21 Solar PV at Qārrat Um as-Saġīr village. (Source: Nahla Nabil)

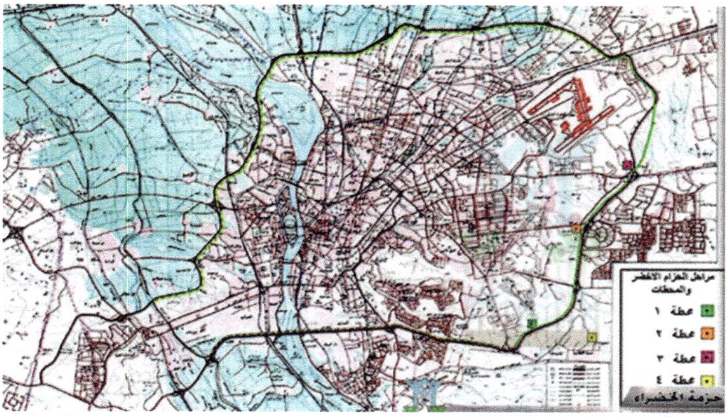

Fig. 6.22 Green belt around Greater Cairo. (Source: EEAA)

which is considered a small number due to the high cost of treating [75]. Moreover, Egypt irrigates about 65,000 trees as part of green belt project shown in Fig. 6.22 around Greater Cairo to improve the water distribution network (pipes, valves, tanks, etc.) to reduce losses caused by leakage [76].

Egypt has a set of experiments in the field using environmentally friendly construction materials, and one of these experiments is the village of New Gourna. It was designed in the 1940s by the architect Hassan Fathy for relocated low-income families [77]. Also, some companies are taking this approach such as Egypt-ecotels that design and build eco-friendly hotels as Qasr El Bagawat Hotel at Kharga Oasis in western desert, which was built from local desert materials [78]. In addition, the National Research Centre in Egypt studies and builds eco-friendly models by recycling agricultural or industrial solid waste and using it in the manufacturing of construction materials or as a compost such as the recycling of chopped rice straw to be used in the manufacture of cement bricks and that of foam and light wire mesh for slabs, beams and stairs [79]. In the 1990s, Egypt built five composting plants, and the first composting plant using locally manufactured equipment was in Zagazig in Sharqiya Governorate, but most of the composting plants do not operate efficiently nor at full capacity [80]. Figure 6.23 shows recycled rooftop project in informal building in Cairo.

In January 2009, Egypt established a national authority for green buildings called Egyptian Green Building Council (EGBC), and in 2010 the council approved a Green Pyramid Rating System (GPRS) for evaluating and rating green building [81].

The rating criteria depend on measurement of sustainable site, water efficiency, energy conservation, saving materials and resources, improving indoor environment quality, design of structure, façade and fenestration, operation and maintenance, reduction of waste, pollution, embodied energy and carbon emissions throughout their life and operating for the benefit of the users and occupants as shown in Table 6.12. However, the code has not been enforced to this day [82]. In 2005, an energy code was issued for improving efficiency in residential and commercial buildings by rating the impact of building envelop, ventilation, heating and lighting [83].

Egypt government in corporation with NGOs and international partners (as European Union and GIZ) replaced and renovated old houses in urban and rural areas to be more resistant against extreme weather, in order to achieve sustainable urban development to alleviate poverty and attain social inclusion; the programme was called 'Participatory Development Programme (PDP) in urban areas'. The target areas of the programme are in Giza Governorate, Masaken Geziret El-Dahab, El-Warraq and Old Boulaq; in Cairo Governorate, Ain Shams, Ezbet El-Nasr and Matariya; and in Qalyubia Governorate, Khosoos, Khanka and Qalyub. The aim of the programme is to assess residents' vulnerabilities and adaptive capacities, raise awareness, share knowledge and experience between partners [84] and implement adaptation measures such as rooftop garden (Fig. 6.24) to increase income, enlarge

Fig. 6.23 Recycled rooftop project, Cairo. (Source: Cairo Climate Talks)

Table 6.12 Green Pyramid Category Weightings (HBRC)

Green pyramid category	Category weighting (%)
Sustainable site, accessibility, ecology	15
Energy efficiency	25
Water efficiency	30
Materials and resources	10
Indoor environmental quality	10
Management	10
Innovation and added value	Bonus

Fig. 6.24 Rooftop garden in a school in Qalyubia. (Source: David Degner)

food affordability and safety, improve local air quality and provide shadowing. In Ezbet El-Nasr settlement, the project implements the following strategies: (1) green wall façade to improve shading of walls and improve air quality, (2) façade plaster and painting to improve solar radiation reflection and (3) adaptation measure implementation which are related to public spaces including green or textile street cover and a community and sidewalk garden [85]. In addition, Egypt and its international partners established the Advisory Board Climate Change in Cities (AB-CCC) for sharing information and knowledge between EEAA, national and local government, academia, development agencies, civil society and private sector on climate change adaptation and mitigation in urban environments [86].

Integrating CC into coastal zone management plan involves relocation and the immediate initiation of the voluntary evacuation of inhabitants in areas at risk of flooding and securing buildings that are expected to be affected by SLR. Measures include protecting and fixing natural sand dune systems that provide natural protection such as enforcing the Mohamed Ali Wall (Fig. 6.25) as a first line of defence of the low lands south of Abu Qir Bay, furthermore reinforcing the northern side of the

Fig. 6.25 Seawall costal area of Alexandria. (Source: Jason Larkin)

international road that runs from Libya to Palestine along the Mediterranean coast to act as a second line of defence to protect the northern zone of the Delta [87].

6.4.2.2 Water Resources Measures

Egypt seeks to use the aquifer as a reservoir to supplement surface water supply during peak periods and recharge during the minimum demand periods, in the assessment of maximum capacity and safe yield of ground water and in the usage of renewable resources to minimize pumping costs through increasing drainage water reuse from about 4.5 BCW/year to 7.0 BCM/year by year 2000 and to 9.0 BCW/year by year 2017. Other measures include improving the distribution network by effective operation and maintenance network and modern technologies such as remote sensing and geographic information systems and harvesting rainfall and flash flood water [88], in addition to using desalination to provide a domestic water supply for some locations along the Mediterranean and the Red Sea coasts as well as in the Sinai Peninsula [89].

6.4.2.3 Ecosystem Measures

Egypt has 30 natural protectorates and intends to proclaim another 14 of marine and wildlife protectorates, shown in Fig. 6.26, to conserve ecosystems, habitats, food chains and individual species [90]. The government also aims to achieve the

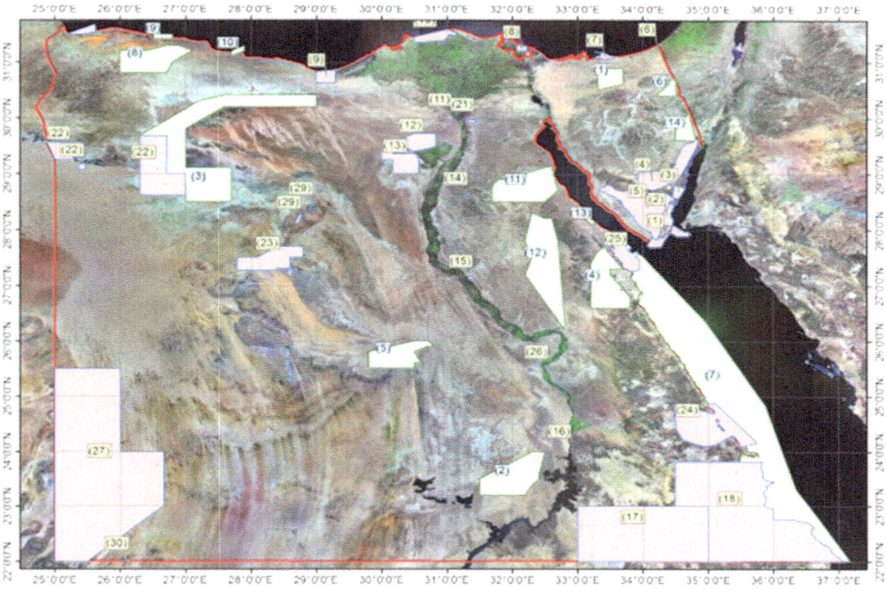

Fig. 6.26 Current (pink) and future (green)

following goals by 2020: (1) investigation and monitoring impact of climate change on ecosystem, (2) restoration of at least 15% of degraded ecosystems, (3) rise of environmental awareness of Egyptians about biodiversity importance, (4) implementation of integrated environmental management systems in touristic sites and the support of strategies by effective legislation and frameworks, (5) encouragement of eco- and sustainable tourism to monitor and control the impact of tourism on biodiversity in vulnerable areas [91] and (6) orientation of tourism growth away from vulnerable areas [33].

It is also imperative to maintain ecosystem functioning and productivity in coastal lagoons and allow habitat restoration by the living shoreline measures such as marsh plantings, beach nourishment and strategically placing structural organic materials for future use. Moreover, consider climate risk aspects into integrated coastal zone management and prevent development that may negatively affect habitats [92]. The United Nations Development Programmes, along with all the aforementioned aspects, concentrate on adaptation to climate change in the Nile Delta through integrated coastal zone management. The programmes started in 2009 to support strengthened institutional and technical capacity of national coastal zone management and to support mainstreaming and implementation of adaptation measures, as well as improving the adaptive capacity of coastal management in pilot projects in each of the three main coastal lakes to be a focus of innovative adaptation activities (Edku, Burullus and Manzala) [93]. Another important goal for the project is the protection of fish resources from infections, diseases and invasive species.

6.4.2.4 Agricultural and Green Spaces Measures

Egypt made improvements towards the use of fertilizer and pesticide, crop rotation and change in planting as well as upgrading the harvesting calendar, modernizing and diversifying agricultural production and cultivation methods. The Egyptian government also took steps in providing information and training to farmers on adaptation measures and raising general awareness about recycling agricultural wastewater, reuse of domestic wastewater in farming and water-saving strategies. Egypt irrigated 30,000 hectares of man-made forest trees with treated wastewater in fourteen governorates to produce timber trees as shown in Fig. 6.27 [94].

Furthermore, there are programmes supported by United Nations which support research activities and build an effective institutional system and simulation exercises on the impacts of climate change on key crops, as well as providing training programmes to improve soil and agricultural land management [95].

Egypt seeks to achieve a restructuring framework of the agricultural insurance system to cover rising risks [13] for livestock, improve the current low productivity cattle and buffalo's breeds and the feeding programmes. These efforts also include improving existing grazing lands, using saline water for forage production, and integrating forages into farming systems [96] as well as relying on biodiversity to improve livestock production, mapping of animal pathogens and its association with CC, and supporting and developing serum and vaccine technologies [33].

Fig. 6.27 Man-made forest trees irrigated by wastewater. (Source: Egyptian Environmental Affairs Agency)

6.4.2.5 Public Health Measures

Egypt aims to identify potential health risks as a result of climate change, raise community awareness about proactive health measures, increase the efficiency of health-care sector and improve the quality of health services [97], as well as developing weather forecast and early warning systems [98].

Table 6.13 investigates the coverage of climate change adaptation measures among sectors including buildings, coastal zones, roads, tourism, health, agriculture, livestock production, fish wealth and water resources and irrigation. The data relies on the official reports of the Ministry of Environment in Egypt and reports of funders and donors of adaptation projects such as the World Bank.

Box 6.4 Overview of Mechanical Wind Water Pumping System and the Possibility of Applying System in Egypt (*Published at International Conference on Improving Sustainability Concept in Developing Hosted by IEREK Held on Grand Nile Tower Hotel December 5–7, 2015*)

Egypt suffers from water stress due to CC impacts; although it possesses a high potential of groundwater availability from different aquifers which cover a large area of Egypt, the absence of infrastructure in rural areas such as water network and electricity grid doubles the pressure on it and leads to the use of fuel-powered pumps for drinking or irrigation. This pollutes the environment and accelerates climate change.

Egypt has high wind potential considered as one of the best wind resources in the world. Wind energy for water pumping is an independent infinite energy resource; it has low running costs and reduces the contribution of greenhouse gases (GHG) to global warming. Perfect meteorological conditions and underground water are available in remote areas, where mechanical wind water pumping system could supply freshwater for drinking, irrigation or industrial production in residential, agricultural and industrial activities in remote or rural areas.

Table 6.13 (a) Climate change adaptation measures in Egypt by sector (1–4)

Sector	Measures
Buildings	Localization of renewable energy [33]
	Rationalization of water and water Recycling [33]
	The use of environmentally friendly construction material [33]

[ ] The red number between two red bracts is reference

Table 6.13 (b) Climate change adaptation measures in Egypt by sector (2–4)

Sector	Measures
Buildings (Cont.)	Recycling of agricultural or industrial solid waste and using this waste in the manufacturing of construction materials or compost [33]
	Issuing a green architecture code. [33] , [21]
	Establishment of a national authority for green buildings using the Leadership in Energy and Environmental Design (LEED) and Building Research Establishment Environmental Assessment Methods (BEREAM) [33]
	Adoption of an energy code [33] , [21]
Coastal zones	Immediately initiate the voluntary evacuation of some inhabitants in the areas at risk of flooding [33] , [21]
	Instigate the replacement and renovation of old houses in urban and rural areas to be more resistant against extreme weather [33]
	Conduct a large-scale campaign to remove all homes located in flood ways [33]
	Secure factories and private and public buildings that are expected to be affected by sea level rise [33]
	Protecting and fixing natural sand dunes systems that provide natural protection. [33] , [21]
	Protecting and enforcing the Mohamed Ali Wall as a first line of defense of the low lands south of Abu-Qir Bay [33] , [21]
	Reinforcing the northern side of the international road along the Mediterranean coast so it can act as a second line of defense for the protection of the northern zone of the Delta [33] , [21]
	Coastal zone management plan [33] , [21] , [13]
Roads	Using martials have resistance to high temperature [33] , [21]
	developed Public transportation [33] , [21]

[ ] The red number between two red bracts is reference

Table 6.13 (c) Climate change adaptation measures in Egypt by sector (3–4)

Sector	Measures
Roads (Cont.)	Redirection of flood ways away from roads [33] , [21]
	protection and diversion dams to direct floods to the main drainage basins [33] , [21] protect desert roads as well as railways running through the desert areas against coverage by sand [33] , [21]
	Review of the road network to determine potential areas vulnerable to flooding in coastal areas [33] , [21]
Tourism	Proclamation of marine and wildlife protectorates [33] , [21]
	Implementation of integrated environmental management systems in touristic sites: [33] , [21]
	Orienting tourism growth away from environmentally sensitive areas and areas that are most at risk to less sensitive and vulnerable ones [33] , [21]
	Forecast warning system [100]
	Analysing the effectiveness of the enforcement of environmental protection laws [33] , [21]
Health	Raising the efficiency of the health care [33] , [21]
	Developing weather and seasonal forecast and early warning systems [33] , [21] Raising awareness on pro-active health measures [33] , [21]
	Promotion of scientific research [33]

[ ] The red number between two red bracts is reference

Table 6.13 (d) Climate change adaptation measures in Egypt by sector (4–4)

Sector	Measures
Agriculture	Building an Effective Institutional System for Crisis and Disaster Management [33]
	Modernization and diversification of agricultural production for food security improvement [33]
	select technologies capable of moderating losses due to heat, frost, drought and new plant pathogens, weeds and pests [33] , [21]
	Soil and Agricultural Land Management [33]
	Restructuring of the agricultural insurance system to cover risks [33] , [21]
	Management of Water Resources and Farm Irrigation [33] , [21]
Livestock Production	Relying on biodiversity to improve livestock production [33]
	Mapping animal pathogens and its association with climate change [33]
	The support and development of serum and vaccine technologies [33]
Fish Wealth	Conservation of biological diversity and reducing the rate of losses [33]
	Protection fish resources from pathological infections and diseases [33]
	Protection of fish communities against new species, which affect the biological and environmental balance [33]
Water Resources and Irrigation	Addition and development of new water sources [33] , [21]
	Improving the distribution network [33]
	Recycling treated wastewater [33]
	Maximum utilization of rainfall and flash flood water [33] , [21]
Water Resources and Irrigation (Cont.)	Using modern techniques in the field like remote sensing and geographic information systems [33]
	Expanded construction of dams and reservoirs to collect this water and use it for drinking or agriculture directly, or for storage in groundwater reservoirs [33]

[ ] The red number between two red bracts is reference

In conclusion, effective climate change adaptation is mapping vulnerability to climate change (CC), forming policies, identifying adaptation measures and then managing all of these by climate change adaptation (CCA) project and going through a wide range of stages ranging from observation, assessment, planning, implementation, monitoring and evaluation to cope with the impacts or consequences of CC in the short or medium term. The next chapter investigates the most important climate change adaptation projects around the world and in Egypt.

Box 6.5 Egypt's National Climate Change Adaptation Measures

1. Urban and coastal areas
   - Use of renewable energy is one of the adjustment measures, but renewable energy consumes a very low rate of power generation in Egypt.
   - Adopt energy code, use environmentally friendly construction material and renewable energy, recycle solid waste and reuse wastewater.
   - Replace or renovate vulnerable building and infrastructure in urban and rural areas to be more resistant against extreme weather.
   - Exploit materials that have resistance to high temperature and depend on eco- transportation.
2. Water
   - Using water recycling and improving distribution network are the most important measures to adapt to water shortage.
3. Agriculture
   - Select technologies capable of moderating losses due to heat, frost and drought and new plant pathogens, weeds and pests.
4. Ecosystem and tourism
   - Implement integrated environmental management systems in touristic sites and marine and wildlife protectorates.
5. Health
   - Promote scientific research and develop early warning systems.

6.5 Integrated Climate Change Policies and Measures: Egypt

Egypt as a developing country is actively addressing issues related to climate change (CC). Egypt also responded early to CC by being one of the first countries to join the collective global efforts that aimed at counterbalancing climate change risks.

Egypt is also one of the first nations signing environmental agreements, including the United Nations Framework Convention on Climate Change (UNFCCC) in 1994 and the Kyoto Protocol in 1994 and ratified in 2005. In 2007, Egypt established the National Committee on Climate Change (NCCC) in order to develop strategies and policies to deal with CC. Recently at the COP 21 in Paris in December 2015, Egypt is one of the signatories of the famous Paris Agreement, and Egypt submits its Climate Action Plan Ahead of 2015 Paris Agreement [100]. In addition, Egypt effectively participated in COP 2016 in Marrakesh and COP 2017 in Bone. Moreover, in 2018, Egypt also presented its Egypt's Voluntary National Review report on July 1, 2018 at the United Nations, and climate change and implementation of SDGs were a solid core of this report, where the SDG 13: Climate Action indicators – *Number of deaths, missing persons and persons affected by disaster per 100,000 people* – was reported at a target 13.1, type 2, at a value of 1.41, year 2017 [101].

From the institutional perspective, Egypt has initiated in 1982 the set-up of the Egyptian Environmental Affairs Agency (EEAA) and launched the Ministry of State for Environmental Affairs in 1997 which became the Ministry of Environment in 2015, and in the same year, Egypt has established the National Council on Climate Change by prime ministerial decree, as a step to centralize policy-making efforts concerning climate change [101].

6.5.1 National Strategy on Climate Change Adaptation

This section focuses on the national and regional strategy on climate change adaptation, if applicable. It includes a description of four main sections:

- The commitments in place at the national level within the international environment (UNFCCC) and the agreements that have been signed for CC
- The national strategy, its goals and commitments, etc.
- The regional strategy, if applicable, based on the national one, and the specificities the region faces
- Paris Agreement – new global climate change agreement, December 2015

On December 12, 2015, *Paris Climate Change Agreement* was approved by the 196 Parties to the United Nations Framework Convention of the Climate Change (UNFCCC) at COP 21 in Paris, France. Egypt signed the Paris Climate Agreement on April 22, 2016, in Paris signature ceremony where 175 participated but has not yet ratified it. 'We respect the experience of others, but we will do it our way. We will ask for your support, but we will look after the well-being of the Egyptian people', the Minister of Environment declared.

On November 11, 2015, Egypt submitted its new climate action plan to the UN Framework Convention on Climate Change (UNFCCC). This Intended Nationally Determined Contribution (INDC) comes in advance of COP 21[110]. The Paris Agreement already entered into force, empowering all countries to prevent average global temperatures rising above 2 °C and to reap the many opportunities that arise from a necessary global transformation to clean and sustainable development. Egypt's INDCs were developed in accordance with Decisions 1/CP.19 and 1/CP.20 of the Arab Republic of Egypt. The INDCs set a series of measures intended to be applied by the country to face climate change in terms of adaptation (to CC impacts) and mitigation (reducing GHG emissions). It includes the following three main elements:

- National circumstances that address general economic conditions, including economic and population growth rates, major sustainable development goals (SDGs) and political circumstances
- National efforts implemented to combat climate change in Egypt, in terms of treating impacts in different sectors (agriculture, water resources, coastal zones, etc.) or efforts made to reduce GHG emissions in different sectors (energy, transportation, industry, etc.)
- Required implementation mechanisms to achieve the objectives of the plan (funding, capacity building and technology transfer) [97]

6.5.2 National Strategy, Goals and Commitments and Sectors Affected

In 2007, Egypt established its National Committee on CC decision in order to develop strategies and policies to deal with CC. As a result, several strategies and policies have been developed, mostly with the support of international donors, including:

- Initial National Communication on Climate Change, Egypt, 1999 – MoEnv and EEAA [102]
- Second National Communication on Climate Change, Egypt, 2010 – EEAA and UNDP [103]
- Third National Communication on Climate Change, Egypt, 2014 – EEAA and UNDP [104]
- National Strategy for Adaptation to Climate Change and Disaster Risk Reduction 2011 (NSACCDRR) – UNDP [33]
- National Air Quality Policy 2015, NAQP – UNEP
- Egypt's Indicators Development
- National Solid Waste Management Policy (NSWMP) 2015 – GIZ

The Egyptian Cabinet – Information and Decision Support Centre (IDSC) – in collaboration with the UNDP has developed *Egypt's National Strategy for Adaptation to Climate Change and Disaster Risk Reduction (NSACCDRR)* in December 2011 [111]. The NSACCDRR's main objective is to increase the flexibility of the Egyptian community when dealing with the risks and disasters that might be caused by climate change and its impact on different sectors and activities. It also aims at strengthening the capacity to absorb and reduce the risks and disasters to be caused by such changes. Also, *Climate Change Risk Management Programme (CCRMP)* was established in 2008 as a result of a partnership between Egypt and the Millennium Development Goals Achievement Fund (MDGAF) [105]. The three main objectives of CCRMP are to (a) integrate GHG mitigation into national policy and investment frameworks; (b) increase CC adaptation capacities, particularly in agriculture and water; and (c) raise awareness regarding the impacts of climate change [105].

In 2010, Egypt published, in response to the SNC to UNFCCC, a *National Environmental, Economic and Development Study (NEEDS)* for CC which outlines the financial and institutional needs for implementing prospective and ongoing adaptation and mitigation measures. This study recognizes that the next phases of CC planning should include a *National Action Plan for Adaptation* (NAPA) and *National Low Carbon Economy Plan (NLCEP)*. The NEEDS report highlights the urgency for developing a GHG monitoring system that aggregates and disseminates information about GHG emissions across sectors. In 2014, the Ministry of Environment (MoEnv) signed an agreement with the Italian counterpart to transform El Gouna City into the first carbon-neutral city in Africa [105].

In addition, the UNEP developed the National Strategy Study entitled: '*Energy Efficiency and Renewable Energy National Study* (EERENS)', which was developed in 2007 as part of both the Mediterranean and National Strategies for Sustainable Development under Priority Field of Action 2: Energy and Climate Change [106, 107]. The key player of this strategy study is the 'Plan Bleu' – a regional activity centre of the UNEP/MAP (Mediterranean Action Plan), created, funded and steered by EC [108]. As far as the CC risks are concerned, Africa is the most vulnerable continent globally, where Egypt is located.

6.5.3 Regional Strategy and the Specificities the Region Faces

The Governorate of Red Sea put forward a strategy to address climate change risks and adaptation. Such strategy is based on the National Strategy on Climate Change. Its main focus is to adapt and reduce risks and disasters caused by climate change, increase awareness of the governorate's staff and enhance its participation aligning with Hurghada city's goals to become the centre of culture and tourism. This strategy will have to be regularly updated for the reason of possible unpredicted changes of climate. The Governorate of Red Sea road map includes strategic objectives, strategic components and actions as follows:

Strategic Objectives

- Define all climate change risks for all sectors.
- Increase the sustainability issues in all sectors.

- Reduce the dependence on fossil fuels.
- Promote renewable energy sources.
- Involve all governorate's and municipality's staff in the effort to transform the city and governorate to be green and sustainable.
- Make the infrastructure of the city of Hurghada and Governorate of Red Sea resilient and a world-class tourism destination.
- Raise awareness and capacity building on CC risks, impacts and adaptation.
- Regularly assess and evaluate the strategy and current situation.

Strategic Components

- Increase the communication with all stakeholders and governorate departments and directorates of all ministries operating in Hurghada.
- Built a vertical communication mechanism to respond to extreme events.
- Promote sustainability and sustainable energy actions in all governorate's institutions and municipal departments to encounter CC risks.
- Develop a database for all changes in the climate in coordination with the National Weather Authority and Hurghada international airport.

Strategic Actions

- Built an observatory in the Hurghada local municipality to monitor the pattern of rainfall and temperature increases and rainfall patterns.
- Give alerts on severe events before it happens in the governorate.
- Construct water barriers in storm areas to collect storm water upon occurring to lessen flood impacts.
- Instal capacity of about 300 kWp solar panels that generate electricity from renewable energy sources to feed in four governorate buildings, in Hurghada.
- Street lighting change from conventional lighting to LED lighting that save about 80% of the energy consumption.

6.5.4 National Climate Change Adaptation and Mitigation Measures

There is a noteworthy legal framework, but nevertheless many policy plans are in implementation, and many government agencies responsible for integrating CC into the national policy agenda were executed. However, the Governorate of Red Sea should develop their local policy and agenda for CCAMM. The climate change mitigation and adaptation strategies are falling under the mandate of the MoEnv and its executive arm EEAA. Also, a climate change unit (CCU) was established in the EEAA in 1992, which has since been upgraded to the Central Department for Climate Change (CDCC) [109].

6.5.5 Climate Change Action Plan (CCAP)

The Climate Change Action Plan (CCAP) has been developed by the MoEnv through the Second National Communication (SNC) to UNFCCC that was established in 2000. An inventory of GHG emissions, including policies, has been created to mitigate and evaluate their economic impacts after reduction. Such policy was not implemented or released until 2015 [57]. Results of the inventory of GHG emission yield an increase of the total emissions at an average of 5.1% annually. The GHG emissions were highly influenced by four sectors that contribute significantly to total emissions. These are (a) energy, which accounted for 61% of the total in 2000, 27% of which was transportation related; (b) agriculture by 16%; (c) industrial processes by 14%; and (d) waste by 9%. Despite the reduction of GHG emissions reported above, there are barriers to GHG mitigation.

Such barriers can be summarized as follows: (a) institutional capacity constraints, (b) lack of information about GHG reduction opportunities and technology and (c) limited access to investment capital needed to execute mitigation technologies and procedures. Nonetheless, an estimate of nearly 8.3 mt CO_2 equivalent reduction due to the existing mitigation efforts focuses on three areas: (a) fuel substitution, (b) renewable energy and (c) energy efficiency [97].

In addition, *Climate Change Risk Management Programme (CCRMP)* was established in 2008 as a result of a partnership between Egypt and the Millennium Development Goals Achievement Fund (MDGAF). The three main objectives are to (I) integrate GHG mitigation into national policy and investment frameworks; (II) increase climate change adaptation capacities, particularly in agriculture and water; and (III) raise awareness regarding the impacts of climate change [57].

Further to the SNC to UNFCCC, Egypt published in 2010 a *National Environmental, Economic and Development Study (NEEDS)* for CC to outline the financial and institutional needs for implementing prospective and ongoing adaptation and mitigation measures. This study recognizes that the next phases of CC planning should include a *National Action Plan for Adaptation* (NAPA) and *National Low Carbon Economy Plan (NLCEP)*. The NEEDS report also highlights the urgency for developing a GHG monitoring system that aggregates and disseminates information about GHG emissions across sectors.

6.6 Conclusions

This chapter examined climate change impact, followed by policies developed and climate adaptation measures taken in developing countries. Egypt, a developing country, is selected to be a reference case as it is one of the most vulnerable countries to CC in Africa based on Chap. 2 conclusions. The case study shows the impact of climate change on the built environment and related urban areas' economics at national scale; in addition, it investigates the adaptation efforts made by the authorities and NGOs. Investigation indicates that Egypt is one of the most vulnerable countries to CC in Africa; as Egypt's climate has already changed, 32 risks were distributed among sectors, one-third of total risks already occurred and rest will occur between 2030 and 2100 according to CC scenarios.

There is evidence on the increase of temperature and sea level rise, in addition to fluctuations in a weather pattern that became more frequent over the past few years. Global warming effect has increased in Egypt since 1960. By 2100, temperature will have increased in Egypt between 3 °C and 3.5 °C, which will accelerate air pollution rate and exacerbates respiratory diseases and mortalities. Moreover, CC will reduce flow of the Nile, increase precipitation that reduce crop production, increase densification and revive dispute over water quotas between the Nile Basin countries. In coastal zones, 1200 km along Mediterranean coast and Nile Delta is exposed to inundation; the scenario projects 276,748 houses and 704 kilometres square to be vulnerable to SLR which threatens two million people and forces most of them to experience displacement. Coral reefs in the Red Sea have undergone significant decline in numbers due to coral bleaching due to change in water temperature and acidity.

According to Organization for Economic Co-operation and Development (OECD), the potential impacts of climate change (CC) on coastal resources are ranked as the most serious risk, followed by that of water resources, agriculture and energy resources as well as urban areas, public health and ecosystem exposure to CC. As a result, Egypt's adaption policies and measures aim at reducing CC associated risks by tools such as mapping vulnerability and monitoring risks along with promoting technical and scientific capacity and developing a participatory adaptation strategy aimed to increase energy and water efficiency, improve land use management, improve public health, retrofit vulnerable buildings and infrastructure and redirect city activities to less vulnerable sites.

Government efforts and actions regarding the national INDC reports, COP 21, COP 22 and COP 23 were presented and discussed. Also, the measures, policies, and action plans for developing country stance towards climate change mitigation and adaptation were depicted. In addition, reference cases in Egypt that highlight such adaptive measures, policies, and strategies as well as the action plan for climate change adaptation were illustrated in the INDC reports.Therefore, effective climate change adaptation is mapping vulnerability to CC, forming policies, identifying adaptation measures and then managing all of these by CCA project and going through a wide range of stages ranging from observation, assessment, planning, implementation, monitoring and evaluation to cope with the impacts or consequences of CC in the short or medium term.

The next chapter evaluates the effectiveness of these measures and policies on the ground through investigating the most important climate change adaptation projects around the world and in Egypt.

References

1. Fulco Ludwig et al, "Climate change impacts on Developing Countries - EU Accountability", The European Parliament's Committee on the Environment, Public Health and Food Safety, 2007.
2. "Hurricane Michael", Guardian News, [Online]. Available at: https://www.theguardian.com/us-news/live/2018/oct/11/hurricane-michael-latest-live-news-updates-florida-weather-evacuations-storm. (Accessed: October 14, 2018).
3. Wade, K., and Jennings, M., "The impact of climate change on the impact of climate change on the global economy", 2016. [Online]. Available at: http://www.schroders.com/en/sysglobalassets/digital/us/pdfs/the-impact-of-climate-change.pdf. (Accessed: September 16, 2018).
4. Aaron Cosbey, "Developing countries interests in climate change action and the implications for a Post-2012 climate change regime", United Nations, New York and Geneva, 2009.
5. Wijaya, A. S., "Climate Change, Global Warming and Global Inequity in Developed and Developing Countries (Analytical Perspective, Issue, Problem and Solution)", in *IOP Conference Series: Earth and Environmental Science*, 2014.
6. E. Ministry of Health, "Egypt says heat wave death toll climbs to 106", sbs News, 19 August 2015. [Online]. Available at: https://www.sbs.com.au/news/egypt-heat-wave-death-toll-rises-to-106. (Accessed: October 14, 2018).
7. Elizabeth Martin Perera and Todd Sanford, "Climate Change and Your Health: Rising Temperatures, Worsening Ozone Pollution", Union of Concerned Scientists, June 2011.
8. David Tresilian, "Egypt and climate change", Ahram weekly, May 2014. [Online]. Available at: weekly.ahram.org.eg/News/6060/32/Egypt-and-climate-change.aspx. (Accessed: April 13, 2016).
9. "Climate: Observations, projections and impacts", UK Met Office, 2011.
10. Joel Smith et al, "Potential Impacts of Climate Change on the Egyptian Economy 2013", United Nations Development Programme, Cairo, Egypt, 2013.
11. C. P. Kumar, "Climate Change and Its Impact on Groundwater Resources", *International Journal of Engineering and Science,* vol. 1, no. 5, pp. 43–60, October 2012.
12. Omran E. Frihy, Khalid M. Dewidar and Mohamed El Raey, "Evaluation of coastal problems at Alexandria, Egypt", *Researchgate,* vol. 30, pp. 281–295, January 1996.
13. Shardul Agrawala et al, "Development and Climate Change in Egypt: Focus on Coastal Resources and the Nile", Organisation for Economic Co-operation and Development, 2004.
14. P. Michael Link, Jasmin Kominek and Jürgen Scheffran, "Impacts of sea level rise on the coastal zones of Egypt", University of Hamburg Research Group Climate Change and Security, 2013.
15. Christopher R. Adams, "Impacts of Temperature Extremes", [Online]. Available at: http://sciencepolicy.colorado.edu/socasp/weather1/adams.html. (Accessed: April 14, 2016).
16. "Historical Climate Variability and Change,Egypt dashboard", The World Bank Group, 2016. [Online]. Available at: http://sdwebx.worldbank.org/climateportalb/home.cfm?page=country_profile&CCode=EGY&ThisTab=RiskOverview. (Accessed: April 17, 2016).
17. Alexandra Fielden Intern, " Ignored Displaced Persons: the plight of IDPs in urban areas", *Policy Development and Evaluation Service, UNHCR,* no. 161, July 2008.
18. "Water Scarcity in Egypt", Ministry of Water Resources and Irrigation, Egypt, February 2014.
19. Bates, B.C., Z.W. Kundzewicz, S. Wu and J.P. Palutiko, "Analyzing regional aspects of climate change and water resources", IPCC Secretariat, June 2008.
20. "EGYPT: Scientists uncertain about climate change impact on Nile, humanitarian news and analysis", IRIN, 2 March 2008. [Online]. Available at: www.irinnews.org/report/77062/egypt-scientists-uncertain-about-climate-change-impact-on-nile. (Accessed: April 15, 2016).
21. Karla Zubrycki et al, "Review of Current and Planned Adaptation Action: North Africa", International Institute for Sustainable Development, November 2011.

22. "World Development Indicators:Freshwater", World Bank, [Online]. Available at: wdi.worldbank.org/table/3.5. (Accessed: April 15, 2016).
23. M. A. Hossain, "Global Warming induced Sea Level Rise on Soil, Land and Crop Production Loss in Bangladesh", in *19th World Congress of Soil Science*, Brisbane, Australia, August 2010.
24. Carolyn Miles et al, "The Urban Disadvantage", Save the Children Federation, 2015.
25. Karla Zubrycki, Alec Crawford, Hilary Hove and Jo-Ellen Parry, "Review of Current and Planned Adaptation Action: North Africa", International Institute for Sustainable Development, November 2011.
26. Mohamed El Raey, "Impacts and Implications of Climate Change for the Coastal Zones of Egypt", The Henry L. Stimson Center, Washington, 2010.
27. Elsharkawy H., Rashed H., and Rached I., "The impacts of SLR on Egypt", 45th ISOCARP Congress 2009, 2009.
28. M. El-Raey, KR. Dewidar and M. El-Hattab, "Mitigation and Adaptation Strategies for Global Change", *Springer,* vol. 4, no. 3, pp. 343–361, September 1999.
29. Ben Haj et al, "vulnerability and impacts of climate change on marine and coastal biological diversity in the Mediterranean Arab Countries", UNEP, Tunis, 2009.
30. "Natural Protectorates Description", Ministry of Environment (MOE),Egyptian Environmental Affairs Agency (EEAA), [Online]. Available at: www.eeaa.gov.eg/en-us/topics/nature/protectorates/protectoratesdescription.aspx. (Accessed: October 14, 2015).
31. T. J. Done, "Phase shifts in coral reef communities and their ecological significance", *Springer,* vol. 247, no. 1, pp. 121–132, November 1992.
32. H. El-Askary et al, "Change detection of coral reef habitat using Landsat-5 TM, Landsat 7 ETM+ and Landsat 8 OLI data in the Red Sea (Hurghada, Egypt)", *International Journal of Remote Sensing,* vol. 35, no. 6, pp. 2327–2346, 2014.
33. "Egypt's National Strategy for Adaptation to Climate Change And Disaster Risk Reduction", The Egyptian cabinet information and decision support center, December 2011.
34. "Egypt tourism numbers to fall less than feared", Reuters Africa, October 2009. [Online]. Available at: https://af.reuters.com/article/investingNews/idAFJOE59J0PG20091020?sp=true. (Accessed: July 21, 2015).
35. "Environment and development in coastal regions and in small islands", International Workshop on Submarine Archaeology and Coastal Management, [Online]. Available at: www.unesco.org/csi/pub/papers2/alex7.htm. (Accessed: July 22, 2015).
36. Ma Bele'n Go'mez Martı'n, "Weather, Climate and Tourism: A Geographical Perspective", Elsevier, Barcelona, Spain, 2005.
37. "Egypt suffers from water poverty", Daily News Egypt, 21 May 2014. [Online]. Available at: www.dailynewsegypt.com/2014/05/21/egypt-suffers-water-poverty-capmas/. (Accessed: April 9, 2016).
38. Hans Günter Brauch, "Climate Change and Conflict Prevention" Can climate change impacts increase conflict potentials", Nature Conservation and Nuclear Safety (BMU), Bonn, June 2002.
39. "Climate Change and Conflict", Federal Ministry for the Environment, Nature Conservation and Nuclear Safety Public Relations Division, 2002.
40. "Egypt – Omayed Biosphere Reserve Project title: Omayed Biosphere Reserve and its surrounding neighbourhood", University of Alexandria, National Committee for the UNESCO-MAB Programme, [Online]. Available at: www.unesco.org/science/doc/mab/Egypt.pdf. (Accessed: July 16, 2016).
41. M. El Raey, "Climate Change and Sustainable Development of Scarce Resources in Egypt", University of Alexandria, Alexandria,Egypt.
42. "World bank site Egypt Dashboard Impacts & Vulnerabilities", World Bank, [Online]. Available at: http://sdwebx.worldbank.org/climateportalb/home.cfm?page=country_profile&CCode=EGY&ThisTab=ImpactsVulnerabilities. (Accessed: April 29, 2015).

43. Climate Guest Contributor, "Top 10 Global Weather Events of 2011", Climate Guest Contributor, 2 JAN 2012. [Online]. Available at: http://thinkprogress.org/climate/2012/01/02/396307/top-10-global-weather-events-of-2011/. (Accessed: March 20, 2016).
44. Ole Mertz et al, "Adaptation to Climate Change in Developing Countries", *Springer,* pp. 743–752, 2009.
45. "Submitted NAPAs", UNFCCC, [Online]. Available at: https://unfccc.int/topics/resilience/workstreams/national-adaptation-programmes-of-action/napas-received. (Accessed: September 15, 2018).
46. Robert Kehew et al, "Addressing Climate Change in National Urban Policy", United Nations Human Settlements Programme (UN-Habitat), 2016.
47. Pelling, M., "Adaptation to Climate Change: From resilience to transformation", Routledge, 2011.
48. Andrea M. Brown, "Uganda's National Urban Policy: The Emerging Response to Poverty, Food Security and Gender in Urban Uganda", The Centre for International Governance Innovation (CIGI), 2013.
49. "Community-Based Flood Preparedness in Hat Yai City", United Nations Framework Convention on Climate Change, [Online]. Available at: https://unfccc.int/climate-action/momentum-for-change/activity-database/momentum-for-change-community-based-flood-preparedness-in-hat-yai-city. (Accessed: September 23, 2018).
50. "Climate Change and Infrastructure in Asian Cities", CAI-Asia and CDIA, Pasig City, Philippines, 2012.
51. "Guide to Climate Change Adaptation in Cities", The International Bank for Reconstruction and Development/The World Bank, 2011.
52. "Adaptation to Climate Change in the Tourism Sector", The World Tourism Organization (UNWTO), 2007.
53. Gina Ziervogel et al, "Climate change and adaptation in African agriculture", Stockholm Environment Institute, 2008.
54. "Climate-Resilient Agriculture in South Asia: An analytical framework and insights from practice", Stockholm Environment Institute, [Online]. Available at: https://www.weadapt.org/knowledge-base/vulnerability/climate-resilient-agriculture-in-south-asia-an-analytical-framework-and-insights-from-practice. (Accessed: September 24, 2018).
55. "Public Health", United Nations Development Program, [Online]. Available at: http://www.adaptation-undp.org/hazards-addressed/public-health. (Accessed: September 24, 2018).
56. "Climate Adaptation to Protect Human Health-UZBEKISTAN", UNDP, [Online]. Available at: http://www.adaptation-undp.org/sites/default/files/downloads/uzbekistan_project_profile.pdf. (Accessed: September 24, 2018).
57. "Climate Change Adaptation and Urban Resilience", [Online]. Available at: egypt-urban.net/climate-change-adaptation-and-urban-resilience/. (Accessed: November 21, 2015).
58. "Climate Change and its Effects on Cities: An Overview from Development Cooperation Perspective", Participatory Development Program in Urban Areas (PDP), Deutsche Gesellschaft für Internationale Zusammenarbeit (GIZ).
59. Dalal Alnaggar, "Water Resources Management and Policies for Egypt", Ministry of Water Resources and Irrigation.
60. "Adapting to climate change induced water stress in the Nile River Basin", The United Nations Environment Programme (UNEP), [Online]. Available at: http://www.unep.org/climatechange/adaptation/EbA/NileRiverBasin/tabid/29584/Default.aspx. (Accessed: April 22, 2016).
61. Daniel Osberghaus and Claudio Baccianti, "Adaptation to Climate Change in the Southern Mediterranean", MEDPRO, February 2013.
62. "Climate change adaptation: Egypt", UNDP-United Nations Development Programme, [Online]. Available at: http://adaptation-undp.org/explore/northern-africa/egypt. (Accessed: April 23, 2016).

63. “Low Emission Capacity Building Project”, UNDP-United Nations Development Programme, 2012.
64. “Urban Solutions Handbook”, WWF, 2017.
65. “UNFCCC”, Vietnam : Building storm-resistance houses, [Online]. Available at: https://unfccc.int/climate-action/momentum-for-change/lighthouse-activities/building-storm-resistant-houses. (Accessed: September 26, 2018).
66. “Adaptation to Coastal Erosion in Vulnerable Areasl Senegal”, United Nations Framework Convention on Climate Change, [Online]. Available at: https://unfccc.int/climate-action/momentum-for-change/lighthouse-activities/adaptation-to-coastal-erosion-in-vulnerable-areas-in-senegal. (Accessed: September 26, 2018).
67. Anne Loftus et al, “Adapting urban water systems to climate change”, ICLEI European Secretariat GmbH, 2011.
68. “Green Coast: community-based coastal restoration in Indonesia, Sri Lanka, Thailand, India and Malaysia”, EBA, [Online]. Available at: http://ebasouth.org/knowledge-centre/resources/casestudy/green-coast-community-based-coastal-restoration-indonesia-sri. (Accessed: September 29, 2018).
69. Quartel, A. Kroon, P.G.E.F. Augustinus and P. Van Santen, “Wave attenuation in coastal mangroves in the Red River Delta, Vietnam”, *Journal of Asian Earth Sciences,* vol. 29, no. 4, pp. 576–584, 2007.
70. Shlomit Paz et al, “Health Aspects of Climate Change in Cities with Mediterranean Climate, and Local Adaptation Plans”, *International Journal of Environmental Research and Public Health,* 2016.
71. “Egypt International energy data and analysis”, U.S. Energy Information Administration (EIA), June 2015.
72. Malina Gont, “EBRD finances solar power plant in Benban”, EBRD, 29 September 2017. [Online]. Available at: https://www.ebrd.com/news/2017/ebrd-finances-solar-power-plant-in-benban.html. (Accessed: October 14, 2018).
73. Nada Badawi, “Tourism Ministry’s calls for Egyptian hotels to ‘go green’ draw skepticism”, Daily News Egypt, June 2, 2013.
74. Omar Hosny, “Tanweer El Heiz; Rural Electrification project”, in *First International Conference on “Solar Energy Solutions for Electricity and Water Supply in Rural Areas*, AUC Cairo, Egypt, 2015.
75. Hussein Abdel-Shafy and Mona S. Mohamed, “Overview on water reuse in Egypt: Present and Future ARTICLE”, *Sustainable Sanitation Practice,* vol. 14, no. 14, pp. 17–25, JANUARY 2013.
76. “Ministry of Environment (MOE)”, Egyptian Environmental Affairs Agency (EEAA) website, [Online]. Available at: www.eeaa.gov.eg/en-us/home.aspx. (Accessed: November 10, 2015).
77. “Earth architecture; Hassan Fathy’s New Gourna”, eartharchitecture, [Online]. Available at: www.eartharchitecture.org/index.php?/categories/23-Egypt. (Accessed: November 13, 2015).
78. “Egyptecotels”, [Online]. Available at: www.egyptecotels.com.eg/index.php?option=com_content&view=article&id=5&Itemid=16#. (Accessed: November 12, 2015).
79. Bakhoum E. S., Garas G. L. and Allam, “Sustainability Analysis of Conventional and Eco-friendly Materials: A Step Towards Green Building”, *ARPN Journal of Engineering and Applied Sciences,* vol. 10, no. 2, FEBRUARY 2015.
80. “Solid Waste Management Strategy for Dakahleya Governorate”, Ministry of Environment (MOE), Egyptian Environmental Affairs Agency (EEAA), March 1999.
81. Karim M. Ayyad and Mostafa Gabr, “Greening Building Codes in Egypt”, in *Sustainable Futures: Architecture and Urbanism in the Global South Kampala*, Uganda, 27–30 June 2012.
82. Mohamed Gamal Ammar, “Evaluation of the Green Egyptian Pyramid”, *Alexandria Engineering Journal,* vol. 51, no. 4, p. 293–304, October 2012.

83. "Egyptian code for enhance energy efficiency in residential building", National Research Centre, Egypt, 2006.
84. "Participatory development programee in urban areas (PDP)", [Online]. Available at: http://egypt-urban.net/. (Accessed: November 15, 2015).
85. Johannes Lückenkötter and Maggie Thabet, "Climate Change, Vulnerability and Adaptation in Selected Urban Areas Overview for the 2nd Stakeholder Workshop", GIZ PDP Cairo, Cairo, 2015.
86. "Memo for the Advisory Board Climate Change in Cities in Egypt (AB-CCC)", [Online]. Available at: http://egypt-urban.net/. (Accessed: November 16, 2015).
87. David Michel and Amit Pandya, "Coastal Zones and Climate Change", The Henry L. Stimson Center, 2010.
88. "Egyptian water resource development", Egyptian National Committee on Irrigation and Drainage (ENCID), [Online]. Available at: http://www.icid.org/v_egypt. (Accessed: November 16, 2015).
89. Samy H. El-Fellaly and Enas Mohammed Abbas Saleh, "Egypt's Experience with Regard to Water Demand Management in Agriculture", in *Eighth International Water Technology Conference*, Alexandria, Egypt, 2004.
90. "Egypt State of the Environment Report 2011", Ministry of Environment,Egypt, 2012.
91. NBSAP 2030 steering committee, "Egypt National Biodiversity Strategy and Action Plan to 203", Egyptian Environmental Affairs Agency (EEAA), 2015.
92. Clare Shelton, "Climate Change Adaptation in Fisheries and Aquaculture", Food and Agriculture Organization of the United Nations, Rome, 2014.
93. The Egyptian Shore Protection Authority, "Adaptation to Climate Change in the Nile Delta through Integrated Coastal Zone Management", United Nations Development Programme, 2009.
94. "Agricultural water productivity as a way of adapting to climate change", Deutsche Gesellschaft für Internationale Zusammenarbeit (GIZ), [Online]. Available at: www.giz.de/en/worldwide/22237.html. (Accessed: November 17, 2015).
95. " Mobilization of the Dahshour World Heritage Site for Community Development", The MDG Achievement Fund (MDG-F), 2013.
96. Mohamed A. El-Nahrawy, "Egypt Pasture/Forage Resource Profile", Food and Agriculture Organization of the United Nations, June 2011.
97. "The Arab Republic of Egypt Intended Nationally Determined Contributions", United Nation Framework Convention on Climate Change, 2015.
98. Karla Zubrycki et al, "Review of Current and Planned Adaptation Action: North Africa: Egypt", International Institute for Sustainable Development, November 2011.
99. "Egypt National Environmental, Economic and Development Study (NEEDS) for Climate Change Under the United Nations Framework Convention on Climate Change", Egyptian Ministry of State for Environmental Affairs, April 2010.
100. "Egypt Submits its Climate Action Plan Ahead of 2015 Paris Agreement", UNFCCC, 11 November 2015. [Online]. Available at: https://unfccc.int/news/egypt-submits-its-climate-action-plan-ahead-of-2015-paris-agreement. (Accessed: October 1, 2018).
101. "Egypt's Voluntary National Review 2018", Ministry of Planning, Monitoring and Administrative Reform, 2018.
102. "Egypt's first National Communication under the United Nations Framework Convention on Climate Change", UNFCCC, 1999.
103. "Egypt's Second Nation Communication under the United Nations Framework Convention on Climate Change", May 2010, UNFCCC.
104. "Egypt's Third Nation Communication under the United Nations Framework Convention on Climate Change", UNFCCC, 2014.
105. Michal Nachmany et al, "Climate Change Legislation in Egypt: The 2015 Global Climate Legislation Study - A Review of Climate Change Legislation in 99 Countries", The Grantham Research Institute on Climate Change and the Environment, 2015.

106. "Guidelines", Ministry of Environment, [Online]. Available at: http://www.eeaa.gov.eg/en-us/mediacenter/reports/guidelines.aspx/. (Accessed: August 10, 2016).
107. "Guidelines of Principles and Procedures for Environmental Impact Assessment", Ministry of Environment, January 2009.
108. "Resilient Cities Acceleration Initiative", UN, 2014. [Online]. Available at: http://www.un.org/climatechange/summit/wp-content/uploads/sites/2/2014/09/RESILIENCE-Resilient-Cities-Acceleration-Initiative.pdf. (Accessed: July 25, 2018).
109. Ministry of Environment- EEAA, [Online]. Available at: www.eeaa.gov.eg/. (Accessed: August 20, 2017).
110. "Egyptain Intended Nationally Determined Contributions", UNFCCC, [Online]. Available at: http://www4.unfccc.int/submissions/INDC/Published%20Documents/Egypt/1/Egyptian%20INDC.pdf. (Accessed: August 24, 2016).
111. "Egypt's National Strategy for Adaptation to Climate Change and Disaster Risk Reduction", NSACCDRR, 2011. [Online]. Available at: http://cairoclimatetalks.net/sites/default/files/Adaptation%20Strategy%20-%20Final%20-%20E.pdf. (Accessed: August 24, 2016).

Chapter 7
Climate Change Adaptation Projects: Global and Developing Countries' Context

Abstract Climate change adaptation through projects globally is essential, particularly in developing countries since these countries are most vulnerable. This chapter examines climate change adaptation (CCA) projects by comparing nine projects in one of the developing countries – Egypt – and global case studies in other developing countries to identify the gaps and track success stories and failures to achieve adaptive learning. In the comparative analysis, nine adaptation projects in Egypt were selected and emphasized on six sectors: energy, urban areas, and awareness raising as well as waste management, ecosystem and agriculture. These sectors were compared to similar projects that were implemented as part of the climate action plans. These projects represent Morocco, Mexico, Brazil, Kiribati, India, the Philippines, Solomon Islands, Central African Republic and Mauritius. The chapter starts with an introduction about climate change adaptation projects and clarifies the scope of study and then discusses and validates the evaluation criteria that depends on seven main features, relevance, effectiveness, sustainability, equity, efficiency, flexibility and stakeholders' participation, and converts them from qualitative to quantitative in order to be measured. The evaluation of the CCA projects includes an overview about each project and identifies the project's description, scope, timeline, aim, challenge and outcome which are required for such evaluation. The results of the comparison were analysed, and the gaps were identified. Finally, the book recommends solutions for overcoming the adaptation action plan short comes.

7.1 Introduction

Climate change has enormous negative risks that threaten the built environment including lives, buildings and infrastructure as well as hinder development around the world. According to Global Humanitarian Forum, 500 million people are at extreme risk due to floods, droughts, landslides and tropical cyclones [1]. As a result, governments, NGOs and private sectors and other stakeholders address climate change adaptation (CCA) through tangible projects and dynamic initiatives that represent practical means to adapt to climate change risks. Monitoring and evaluating process is responsible for ensuring that CCA projects are able to meet

M. M. Aboulnaga et al., *Urban Climate Change Adaptation in Developing Countries*, https://doi.org/10.1007/978-3-030-05405-2_7

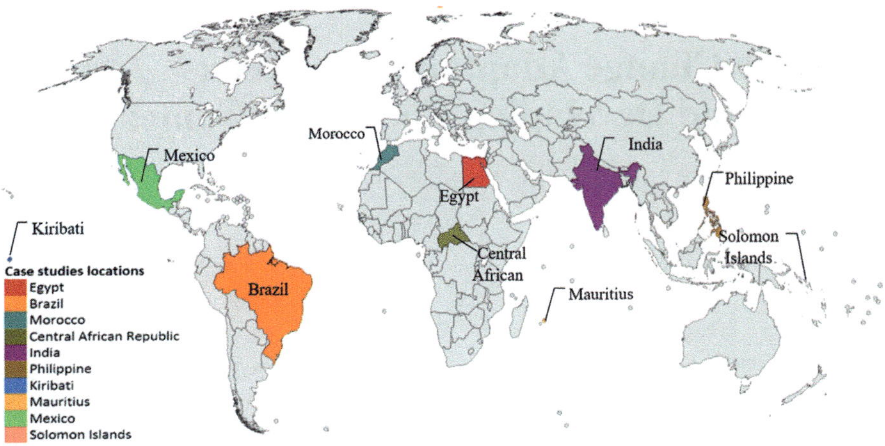

Fig. 7.1 Climate change adaptation case study locations. (Source: Developed by authors)

their objectives, targets and deliverables through tracking success stories and failures in order to strengthen knowledge about the impacts of these CCA projects on mitigating such risks.

The chapter evaluates the performance of Egypt's CCA in comparison to that of global projects in other developing countries to investigate gaps and extract lessons learned. The chapter also discusses climate change adaptation projects which were implemented as part of the official CCA policy. The selected case studies have the same adaptation theme and face almost the same challenges. All of the case studies were selected from developing countries because they have approximately common social, economic and environmental challenges that make the comparison fair.

The study investigates project scope, timeline, goals and contractor as well as total cost, partners, challenges and outcomes for 20 projects, 9 of which are implemented in Egypt and 9 are implemented in other developing countries from three continents including Brazil and Mexico (Central and Latin America); India, the Philippines and Solomon Islands (Asia); Kiribati (North Pacific Ocean); and Central African Republic, Mauritius and Morocco (Africa) as illustrated in Fig. 7.1.

7.2 Evaluation Criteria

The criteria of evaluation are used to assess the effectiveness of an adaptation intervention and avoid economic, environmental and social side effects [2]. Many studies discuss evaluation criteria in order to develop appropriate approaches to be used for assessment to support climate change adaptation (CCA) and identify main gaps, while assessing adaptation interventions. Examples of adaptation studies were reviewed and classified into two types: first type is supported by international institution whereas the second type is carried out by individual experts. For the first type, there are two well-known studies: (a) Merylyn Mckenzie Hedger et al. (August 2008), which is supported by global financier – the Global Environment Facility (GEF); and (b) Schipper L., Liu W., Krawanchid D., and Chanthy S. (2010), which is supported by Intergovernmental body of Mekong River Commission. An example of the second type of studies that are carried out by individual experts is Stelios Grafakos and Veronica Olivotto (March 2013). Nonetheless, other studies are shown in Table 7.1. These studies represent a sample of approaches' listing, frameworks and set criteria used for the adaptation evaluation.

Through investigation, it was realized that these climate change adaptation project (CCAP) evaluation approaches have common evaluation criteria such as efficiency, effectiveness, equity, stakeholders' precipitation, relevance, sustainability and flexibility. Moreover, financiers add coherence to the criteria sets to be sure that the project does not conflict with other adaptation efforts and planned policies. The study conducted by Sabine Weiland and Jenny Tröltzsch et al., (June 2015), called 'BASE Evaluation Criteria for Climate Adaptation' or BECCA, is based on a comprehensive review of 57 documents on climate adaptation evaluation covering academic literature and CCA world policies that demonstrate the aforementioned criteria.

The study conducted in this book followed a similar evaluation approach and depended on seven main criteria points to weigh and determine successful adaptation. The seven criteria are (a) relevance, (b) effectiveness, (c) sustainability, (d) equity, (e) efficiency, (f) flexibility and (g) stakeholders' participation. Successful adaptation is the adaptation which succeeds in dealing with the following seven criteria:

(i) *Relevance*: It indicates the extent of how far the adaptation project measures address climate change impact.
(ii) *Effectiveness*: It is the adaptation intervention that achieves its stated objectives, reduces vulnerability, increases adaptive capacity and enhances level of protection.
(iii) *Sustainability*: The sustainability of an adaptation is concerned with looking beyond the immediate effect of the intervention's impact. It considers the longer-term benefits and environmental, social and economic impacts. Sustainable adaptation includes strong elements of partnership building, community engagement, education and awareness raising, mainstreamed into existing development processes to offer 'win-win' solutions [3].
(iv) *Equity*: It evaluates dimensions of inequality in adaptation because inequitable adaptation interventions will undermine the potential for welfare gains in the future, so adaptation should respect equity by considering equalities between sectors, regions and societies [4].
(v) *Efficiency (cost-effectiveness)*: It is used to compare the costs of alternative ways of producing the same or similar results to assess the least cost path to reach a given target and address agreed acceptable risk levels.
(vi) *Flexibility*: It evaluates how the CC adaptation deals with future uncertainty. It also means how adaptation is dealing with a wide range of CC projects under

Table 7.1 Sample of studies that discussed the evaluation criteria of CCA projects

Study	Recommended criteria	Author	Sponsor	Citation
Global Climate Change Alliance: Pacific Small Island States Individual Country Evaluation Report – Tonga	Relevance Coherence Effectiveness Efficiency Sustainability Equity Stakeholders' participation	Pacific Research and Evaluation Associates	International development organization	Pacific Research and Evaluation Associates. (May 2016) Global Climate Change Alliance: Pacific Small Island States Individual Country Evaluation Report – Tonga, Pacific Community
Evaluation Framework	Relevance Effectiveness Efficiency Sustainability Coherence Equity Stakeholders' participation	Adaptation Fund Board	The Adaptation Fund	Adaptation Fund Board. Evaluation Framework. The Adaptation Fund
Evaluation of Adaptation to Climate Change from a Development Perspective	Effectiveness Flexibility Equity Efficiency Sustainability	Merylyn Mckenzie Hedger et al., Institute of Development Studies	GEF Evaluation Office	Merylyn Mckenzie Hedger et al. (August 2008) Evaluation of Adaptation to Climate Change From a Development Perspective. A study commissioned by the GEF Evaluation Office and financed by DFID
Towards an Integrated Evaluation Framework of Climate Change Adaptation Projects	Relevance Effectiveness Efficiency Equity	Stelios Grafakos and Veronica Olivotto	The Institute for Housing and Urban Development Studies (IHS)	Stelios Grafakos and Veronica Olivotto (2013) Towards an Integrated Evaluation Framework of Climate Change Adaptation Projects, European Climate Change Adaptation conference March 18–20 2013
Glossary of Key Terms in Evaluation and Results Based Management	Relevance Efficiency Effectiveness Sustainability	Niels Dabelstein	Organization for Economic Co-operation and Development (OECD)	Niels Dabelstein (2010) Glossary of Key Terms in Evaluation and Results Based Management, Organisation for Economic Co-operation and Development (OECD)

Table 7.1 (continued)

Study	Recommended criteria	Author	Sponsor	Citation
Evaluating and prioritizing technologies for adaptation to climate change. A hands-on guidance to multi criteria analysis (MCA) and the identification and assessment of related criteria	Relevant Coherence Sustainability Effectiveness Stakeholders' participation Efficiency	Traerup, Sara Laerke Meltofte; Bakkegaard, Riyong Kim	UNEP DTU Partnership	Traerup, S. L. M. and Bakkegaard, R. K. (2015). Evaluating and prioritizing technologies for adaptation to climate change. A hands-on guidance to multi criteria analysis (MCA) and the identification and assessment of related criteria. Copenhagen: UNEP DTU Partnership
BASE Evaluation Criteria for Climate Adaptation (BECCA)	Effectiveness Efficiency Equity Sustainability Coherence Stakeholders' participation Flexibility	Sabine Weiland and Jenny Tröltzsch.	European Union's programmes for research	Sabine Weiland and Jenny Tröltzsch (2015) BASE Evaluation Criteria for Climate Adaptation (BECCA) European Union's programmes for research

different emissions' scenarios since response to worst-case scenario will be safer, but nevertheless, it will be extremely expensive and reduce available resources. Therefore, successful adaptation improves current climate resilience and depends on 'no regret' and 'win-win' interventions that allow for better decision downstream.

(vii) *Stakeholders' participation*: It is achieved by increasing awareness among key stakeholders on the risk that climate change poses to the achievement of national development priorities. This engagement involves multiple stakeholders from different sectors, subnational governments and civil societies, to strengthen the policy design and instruments, build commitment to implement adaptation actions and ensure that policies are informed by practical ground-level experience [5].

The seven evaluation criteria are qualitative and therefore cannot be measured in numbers. Hence, these criteria were given weight on a scale from 0 to 1; the scale '0' means the project does not meet the criteria, whereas the scale '1' means the project is doing well in this regard, and the scale '0.5' means the project has an average performance. This scale is shown in Table 7.2. In order to make the criteria quantitative, the book followed a specific weight of each criterion depending on certain dimensions, which are discussed in the following section.

Table 7.2 Indication of evaluation criteria performance: a numerical scale

Numerical scale	Evaluation criteria performance
1	The project is doing well in this regard
0.5	The project has an average performance
0	The project does not meet the criteria

7.2.1 Converting Criteria Sets to Be Quantitative

First, the relevance indicates the extent to how far the adaptation project measures addressed CC impact; therefore, if the project measures are not related to CC adaptation, it scores 0 points, indirectly relevant project scores 0.5 points and directly relevant scores 1 point, all of project measures assessed collectively.

Second, sustainability evaluates the performance of a project in enhancing economy, social and environmental condition as well as supporting SDGs. Sustainable projects that enhance economy, social and environmental condition score 1 point, while projects which do not have any negative effect on economy, social and environmental condition score 0.5 and unsustainable projects score 0.0.

Third, effectiveness measures the degree a project can cope with adverse effects of climate change through reduced exposure or sensitivity and increased adaptive capacity. Exposure refers to the degree of exposing to significant climatic variations, while sensitivity refers to the degree to which a system is adversely affected by climatic change, as well as adaptive capacity measures the ability of a system to adjust to climate change, moderate risks, take advantage of opportunities and to cope with the consequences [6]. Low effectiveness project which has a 50% success rate of achieving its objectives scores 0.0 points, while moderate effectiveness project that has a 50–75% success rate of achieving its objectives scores 0.5 points and high effectiveness that has a 75 or more percent success rate of achieving its objectives scores 1 point.

Fourth, equity evaluates the level to which a project considers equalities between sectors, regions and societies among proportions of beneficiaries from CCA project and the level of attention to the most vulnerable groups by involving them in planning, prioritization, monitoring and evaluation [7]. CCA projects that respect equity through involving NGOs and local communities in decision-making and do not limit benefit to a specific group score 1 point, while projects considering equalities between societies among proportion of beneficiaries but do not involve them in decision-making score 0.5, and adaptation projects that have inequality through limiting decision-making to nation-states or limiting benefit to a specific group score 0.

Fifth, efficiency evaluates economic viability of adaptation measures through comparing between the benefits gained and costs of implementation [2]. Projects are considered cost-efficient if they bring higher benefits in comparison to costs of implementation. If an actual project cost is higher than the budgeted cost, which means negative budget variances, the project scores 0 point, while if a project cost is equal to or lower than the baseline budget, which means positive budget variances, the project scores 1 point.

Sixth, the flexibility of adaptation measure refers to its ability to adjust under different CC scenarios and its self-organization to adapt to stress, change and deal with uncertainty [8]. Flexible measure that has positive effects reached without climate change risk (CCR) scores 1 point, semi-flexible measure that has positive effect reached under different scenario scores 0.5, and non-flexible measure scores 0.0 points.

Finally, the scope of stakeholders' participation in the adaptation is divided to a top-down approach, a bottom-up approach or a lateral approach. In the top-down approach, adaptation planning involves government level. Contrariwise, the bottom-up approach empowers communities in all stages of adaptation, and lateral approach combines between two approaches by collaborative arrangement between government and stakeholders in making decision [9]. The top-down scores 0.0 point, whereas the bottom-up scores 0.5 and the lateral approach scores 1 point.

Table 7.3 presents the rating performance of climate change adaptation projects (CCAP) according to the overall evaluation criteria score. The scale is ranging from 0 to 7. A score range between 0 and 4.9 means that the project's performance is weak, whereas a score range between 5 and 6.3 means moderate project's performance; and a score range between 6.4 and 7 indicates strong project's performance. This detailed scale describes each range according to the criteria used by the World Bank to evaluate performance although the rating values are adopted by the author according to selected criteria, and it is divided and presented in Table 7.3 [10]. The scale of the highly unsatisfactory rating ranges between 0.0 and 2.66; an unsatisfactory rating ranges between 2.67 to 4.2, whereas the moderately unsatisfactory rating ranges between 4.3 and 4.9. The satisfactory, moderately satisfactory and highly satisfactory rating ranges from 5.7 to 6.3, 5 to 5.6 and 6.4 to 7.

After presenting and discussing the evaluation criteria in this section, the next part is depicting the evaluation of climate change adaptation projects (CCAP) and the case studies in developing countries. It is also comparing these projects to existing CCAPs in Egypt. This assessment is carried out mainly to identify the gap, track adaptation progress, and examine CCAPs' evaluation criteria to weigh the performance of these adaptation projects. In order to comprehend the next section, it is imperative to provide the main points to understand the following terms:

- *Project scope*: refers to the project objective of CC mitigation, adaptation or both
- *Timeline*: is a time scale of the project, which shows the start date and end date

Table 7.3 Rating performance of CCAPs based on evaluation criteria score

Rating *(as per the WB)*	Overall evaluation criteria score
Highly satisfactory	6.4–7
Satisfactory	5.7–6.3
Moderately satisfactory	5–5.6
Moderately unsatisfactory	4.3–4.9
Unsatisfactory	2.67–4.2
Highly unsatisfactory	0–2.66

- *Goals*: are the desired results that the project envisions to be achieved at the end of the timeline or during any specified period in the process for overcoming the obstacles and keeping track of progress in relation to the targets – in some projects, goals are divided into major and minor
- *Contractor*: is company or an agency responsible for project's implementation
- *Total cost*: is the sum of project's implementation cost, consultancy services and monitoring services, plus additional expenses, such as compensation of affected groups by the project
- *Partner*: is an organization or an agency, which provides funding or logistical support through lending a hand in the form of providing technical information, advisory services or other issues essential to the proper operation of the project
- *Challenge*: refers to obstacles that can hinder achievement of the project goals or problems that might increase the vulnerability of a sector to climate change
- *Outcomes*: means the benefits that affect the sector or the environment that result from the implementation of adaptation projects, thus increasing their resilience and mitigating vulnerability to climate change.

7.3 Evaluation of Climate Change Adaptation Projects Globally: Comparison of Developing Countries Versus Egypt

Many projects were implemented worldwide in an effort to climate change adaptation. The performance of climate change adaptation projects (CCAPs) was evaluated by comparing nine CCAPs in one of the developing countries – Egypt – with global CCAP case studies in developing countries to identify the gaps and track success stories and failures to achieve adaptive learning.

The nine climate change adaptation projects (CCAPs) focus on six sectors, which include, energy, urban areas, awareness raising, waste management, ecosystem and agriculture. These six sectors were compared with similar CCAPs implemented, as part of climate action plan, in various developing countries worldwide, including Brazil and Mexico (Central and Latin America); India, the Philippines, and Solomon Islands (Asia); Kiribati (North Pacific Ocean); and Central African Republic, Mauritius, and Morocco (Africa).

An overview about each project's description, scope and timeline as well as the aims, challenges and outcomes was presented and discussed. It forms an important path for evaluation and comparison. It also compares the climate change adaptation projects (CCAPs) that have the similar adaptation themes through evaluating the projects' performance based on seven main criteria: (i) relevance, (ii) effectiveness, (iii) sustainability, (iv) equity, (v) efficiency, (vi) flexibility and (vii) stakeholders' participation. All of these projects are depending on quantitative assessment.

Figure 7.2 and the following tables present the climate change adaptation projects (CCAPs) in six sectors: (a) energy, (b) urban areas, (c) awareness raising, (d) waste management, (e) ecosystem and (f) agriculture.

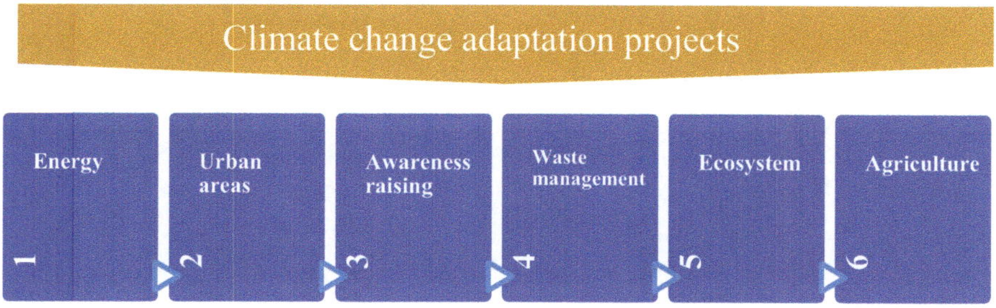

Fig. 7.2 Climate change adaptation projects' sectors. (Source: Developed by authors)

It is imperative to state that the selection of these projects was based on the three main points: (a) the project's size; (b) similarity in each of the above-mentioned six sectors in the selected developing countries; and (c) the project has the same criteria.

7.3.1 Climate Change Adaptation Projects in Energy Sector (Case Studies) (Figs. 7.3, 7.4, 7.5, and 7.6)

Egypt	Kureimat Solar Thermal Hybrid Project
Description	It is a power plant that produces 140 MW by combined cycle and a solar power of about 61 MW It depends on solar thermal energy to steam water by solar collectors to power a thermodynamic cycle
Scope	Climate change mitigation and adaptation
Timeline	December 2007 to October 2011
Contractor	Orascom Construction with Solar Millennium
Total cost	327 million US$
Partner	World Bank and Japan Bank for International Cooperation New and Renewable Energy Authority (NREA)
Aim	Increase share of solar power by 20 MW output Reduce CO_2 emissions from energy generation
Challenge	Electricity demand rapid by 6.5 percent per year Energy shortage and air pollution
Outcome	Increasing share of solar energy by 20 megawatts Avoidance of 40,000 tons of CO_2 emissions Build up capacity of technicians and engineers of NREA (New and Renewable Energy Authority)
Sources	[11–14]

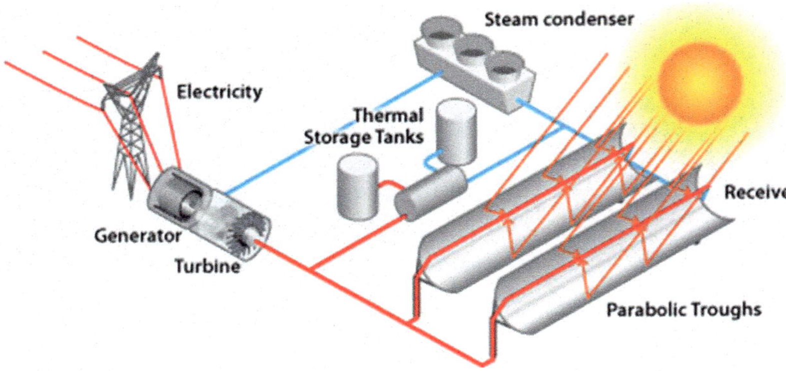

Fig. 7.3 Hybrid solar thermal power plant component. (Source: US Department of Energy)

 Morocco	Integrated Solar Combined Cycle Power Project
Description	The project is a solar/gas hybrid power station that generates about 470 MW, of which 450 MW from combined cycle gas turbine and 20 MW from the solar components The solar output is estimated at 1.13% of the annual production The project includes construction of transmission lines, substation, access roads, gas pipeline and environmental and social development management
Scope	Climate change mitigation and adaptation
Timeline	April 2007 to December 2012
Contractor	Office National De L'electricite (ONE)
Total cost	567 million US$
Partner	African Development Bank Global Environment Facility (GEF)
Aim	Increase electricity capacity to cope growth in electricity demand Promote renewable energy and energy efficiency Reduce greenhouse gas emissions by 610,000 tons over the 25 years (economic life of this plant)
Challenge	Involuntary resettlement Energy shortage due to growing demand Lack of scientific capacity High costs of technology
Outcome	Increasing market share of low greenhouse gas emitting technologies Reduce air pollutant emissions Increase efficiency of ONEE staff
Sources	[15, 16]

Fig. 7.4 Integrated solar combined cycle power project. (Source: Abeinsa Company)

Evaluation

The two projects aim at coping with adverse effects of the increase in energy demand due to global warming and mitigating GHG. This makes these projects very relevant; hence they score 1. Both projects depend on top-bottom approach in decision-making and mentoring; hence community participation in the process is weak, which in turn led to a score of 0 (Table 7.4). Renewable energy projects that are characterized by positive effects withstand climate change risk (CCR) and are doing well under different CC scenarios. In addition, the two projects succeeded in achieving more than 75% of their set targets through mitigating the GHG emissions. Also, these projects increased the share of solar energy, thus led to be evaluated with score 1 in terms of the effectiveness and flexibility criteria (Table 7.4). Moreover, both projects are currently providing electricity to a large number of households, hence contributing to better living standards and supporting economic growth. Kureimat project scores 1 in sustainability, whereas Morocco project scores 0 due to involuntary resettlement of local community. Although the Kureimat project has a small negative impact on environment due to discharging cooling water in the Nile River, it has been remedied by the treatment of water before discharge. Furthermore, the two projects consider equalities between societies among proportion of beneficiaries, but did not involve them in decision-making; hence both score 0.5 point in equity. Projects' cost is equal to the baseline budget, which means 'positive budget variance projects' resulting in recording a perfect score in efficiency. Finally, the overall evaluation of Kureimat project is moderately satisfactory, whereas the Morocco project is moderately unsatisfactory due to weak stakeholders' participation in the project and non-sustainability approach as well.

Table 7.4 Weight evaluation criteria of Kureimat and CCP projects

Evaluation criteria	Kureimat Solar Thermal Hybrid Project  Egypt	Integrated Solar Combined Cycle Power Project  Morocco
Relevance	1	1
Stakeholder's participation	0	0
Flexibility	1	1
Effectiveness	1	1
Sustainability	1	0
Equity	0.5	0.5
Efficiency	1	1
Total	5.5	4.5

 Egypt	Wind Power Development Project
Description	The project is located at the Gulf of Suez that has enormous wind resources with a wind speed 7–12 m/s The project generates 250 MW through wind turbines The project comprises transmission infrastructure development with technical assistance
Scope	Climate change mitigation and adaptation
Timeline	June 15, 2010, to December 31, 2017
Contractor	Egyptian Electricity Transmission Company – MoEE
Total cost	796 million US$
Partner	European Investment Bank (EIB) European Bank for Reconstruction and Development (EBRD) Clean Technology Fund (CTF) and others
Aim	Develop business models for scaling-up wind power in Egypt Transmission infrastructure Technical assistance
Challenge	Growing electric energy demand Low capacity of technicians Involuntary resettlement
Outcome	Pave the way to increase wind generation to 12% of expected generating electricity in 2020 Reduce GHG emission by ~1.5 million tons CO_2 First private sector investment in wind power Increase energy transmission efficiency Build up capacity of technicians and engineers of NREA (New and Renewable Energy Authority)
Sources	[17, 18]

Fig. 7.5 Wind farms in Gulf of Suez, Egypt. (Source: Creative Commons)

Mexico	Wind Umbrella Project
Description	The project, which is considered the first large-scale wind energy plant in Mexico, consists of 98 wind turbines and is generating 340 GWh annually. The project's associated interconnection system is owned and operated by CFE. Annual reduction of GHG of about 200 k tons of CO_2.
Scope	Climate change mitigation and adaptation
Timeline	December 12, 2006 to December 31, 2016
Contractor	The Federal Electricity Commission (CFE)
Total cost	72 million US$
Partner	European Bank for Reconstruction and Development (EBRD)
Aim	Reduce GHG from power generation in Mexico Promote investment in wind energy Diversify the sources of power generation in the country
Challenge	Collision birds' migration with wind turbine blades Project located in land of indigenous people
Outcome	Development of the international carbon market Reduce about four million CO_2 emission over 20 years (operation period)
Sources	[19, 20]

Evaluation

In most of the renewable energy projects, the aim of both projects is to cope with the adverse effect of energy demand increase due to global warming and urban heat island (UHI) effect and mitigate GHG, which make these projects very relevant to CC and get the high score in the evaluation criterion – Relevance (Table 7.5). Egypt's wind project scores 0 in the criterion – Stakeholders' participation – due to the fact that the project depends on a top-bottom approach in decision-making;

Fig. 7.6 La Venta wind farm in Oaxaca, Mexico. (Source: Difer-wikicommons)

nevertheless, the project was built in a wide desert area with almost very low habitation around it, whereas Mexico's Wind Umbrella Project employed specialists to work with the local community in order to agree on a fair compensation for the use of land and educate them about the benefits of wind energy in the region; hence this project scores 1. The projects' positive effects were reached without climate change risk (CCR); hence both projects get 1 in Flexibility criterion, but the Mexico's Wind Umbrella Project succeeded in achieving more than 75% of the target by mitigating emissions, whereas Egypt's wind power project was delayed due to political situation after January 2011; nevertheless, the authorities pledged to finish it on schedule. As a result, Mexico project scored fully in Effectiveness criterion, but Egypt's project did not. Although such project is contributing to better living standards and economic growth, both projects are in close proximity to birds' migration routes, and for this reason, both projects score 0 in Sustainability criterion. Prior to the initiation of the Mexico's project, the authorities provided public lighting, paved the streets, and computers for secondary schools. During the project operation, authorities gave priority to jobs' opportunities and paid the land rent to local community, without compromising the traditional uses in the area. All the above resulted in a perfect score in terms of the evaluation of the Equity criterion. For Egypt's project, it is considered to provide equalities among beneficiaries, but it does not involve them in the decision-making process; thus it scores 0.5 point in the evaluation regarding the Equity criterion. Nonetheless, both projects score 1 point (highest) in the Efficiency criterion (Table 7.5). This is due to their projects' costs, which are equal to the baseline budget, meaning positive budget variances. The overall evaluation of Mexico' project is satisfactory, whereas Egypt's project is unsatisfactory due to weak performance of stakeholders' participation, sustainability, and the delay in the execution timetable.

Table 7.5 Weight evaluation criteria of Egypt and Mexico: wind projects

Evaluation criteria	Wind Power Development Project  Egypt	Wind Umbrella Project  Mexico
Relevance	1	1
Stakeholder's participation	0	1
Flexibility	1	1
Effectiveness	0	1
Sustainability	0	0
Equity	0.5	1
Efficiency	1	1
Total	3.5	6

7.3.2 *CCA Projects' Case Studies in the Waste Management Sector (Figs. 7.7 and 7.8)*

 Egypt	ONYX Solid Waste Alexandria – Carbon
Description	The Landfill Gas Flaring CDM Project of Borg El Arab aims at upgrading the current landfill in Borg Al-Arab, Alexandria in accordance to the environmental standard and regulations. Also, a new extension part was constructed, as a waste management system (landfill extract LFG), to use on-site evaporator or flared.
Scope	Climate change mitigation and adaptation
Timeline	February 19, 2006, to December 31, 2015
Contractor	ONYX Egypt
Total cost	N/A
Partner	World Bank (WD)
Aim	Sizing landfill to environmental standard and regulations Reducing fugitive methane emissions Develop awareness raising of workers Green belt of trees around the landfill site to act as a noise barrier
Challenge	Soil erosion and land degradation Contamination of underground water from leachate Increase of pest species
Outcome	Reduction of 75% in pollution load by industrial facilities Perfect solid waste management, pollution management and environmental health Mitigation measures and treatment technologies are very suitable for pollution control Improvement of public health by reducing GHG emission by 3.7 Creation of 4500 job opportunities
Sources	[21–23]

Fig. 7.7 ONYX solid waste landfill. (Source: United Nations Framework Convention on Climate Change)

 Brazil	BR Nova Gerar Carbon Finance and Solid Waste Management Project II (Brazil)
Description	The project is linked to the integrated solid waste management of Brazil It comprised five individual subprojects The project aims at rehabilitating and closing open dump sites and constructs eco- landfill that depends on LFG capture and management. It also aims at generating flaring methane and using it in electricity generation or selling. The project alleviates poverty and improves social conditions of waste pickers and surrounding populations
Scope	Climate change mitigation and adaptation
Timeline	December 3, 2008, to June 30, 2015
Contractor	Nova Gerar Eco Energia Ltda.
Total cost	10 million US$
Partner	International Bank for Reconstruction and Development Spanish Carbon Fund
Aim	Mitigate greenhouse gas emissions Reduce environmental and social impacts of solid waste disposal
Challenge	One of the landfills located near wetlands
Outcome	Sell emission reductions (ERS) from clean development mechanism (CDM) projects according to the Kyoto Protocol Sustainable development for waste picker communities Dealing with environmental impacts of dumpsites
Sources	[24]

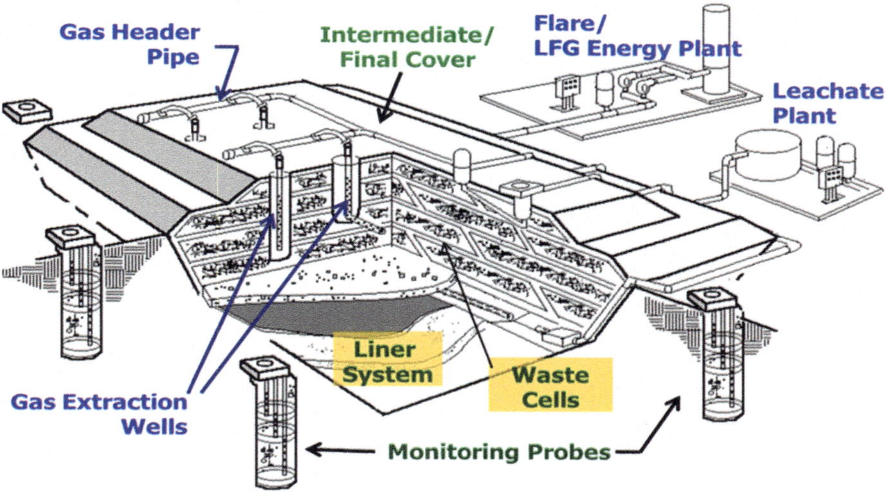

Fig. 7.8 Landfill gas collection system. (Source: US EPA – LMOP)

Evaluation

The two projects are part of solid waste management which is one of the CCA measures; they also mitigate GHG which makes them very relevant to CC adaptation and mitigation so both projects score 1. ONYX project involves only local authorities and neglects waste pickers, so it gets the lowest score in stakeholders' participation; contrariwise, Novagerar project's initial comprehensive framework is to engage waste pickers and local stakeholders to determine their priorities and concern, so the project gets a score of 1.

Both projects enhance environmental, social and economic condition. The projects' success in reducing exposure and sensitivity and increasing adaptive capacity under different scenarios allow them to get highest score in Sustainability, Flexibility and Effectiveness criteria. ONYX project considers equalities between beneficiaries but limit decision-making to nation-states, so it gets a moderate score in Equity criteria, in contrast with Novagerar project which get full score as shown in Table 7.6 due to improving living and working conditions and providing alternatives to waste pickers living on the landfill, as well as involving them in planning and monitoring. Both projects are cost-effective and provide higher benefits in comparison to costs of implementation so they are efficient. The overall evaluation of ONYX project is moderately satisfactory due to absence of stakeholders' involvement and that of Novagerar is highly satisfactory.

Table 7.6 Weight evaluation criteria of ONYX and Novagerar projects

Closed Evaluation criteria	ONYX Solid Waste Alexandria – Carbon Project  Egypt	Novagerar Carbon Finance and Solid Waste Management Project II  Brazil
Relevance	1	1
Stakeholders' participation	0	1
Flexibility	1	1
Effectiveness	1	1
Sustainability	1	1
Equity	0.5	1
Efficiency	1	1
Total	5.5	7

7.3.3 CCA Projects' Case Studies in Urban Areas Sector (Figs. 7.9, 7.10, 7.11, and 7.12)

 Egypt	Mohamed Ali Seawall Project
Description	Built in 1830 to protect low-lying levels varying between −1.5 and −2.5 below mean sea level at Abu Qir Bay southeast of Alexandria The narrow strips of elevated land protect the southern area of Port Said against inundation by sea level rise (SLR) which was repaired and upgraded in 1981 and 2009 and after the surge storm in 2010
Scope	Climate change adaptation
Timeline	1830 – till present
Contractor	Egyptian General Authority for Coastal Protection (EGACP)
Total cost	N/A
Partner	N/A
Aim	First line of defence against SLR Protect vulnerable settlements and infrastructure
Challenge	Lack of funds (international funders not prefer hard measure) Environmental side effects (loss of intertidal habitat and sandy beaches High maintenance cost
Outcome	Play main role in the protection of the lowland area Protect shoreline and coastal roads against coastal erosion
Sources	[25, 26]

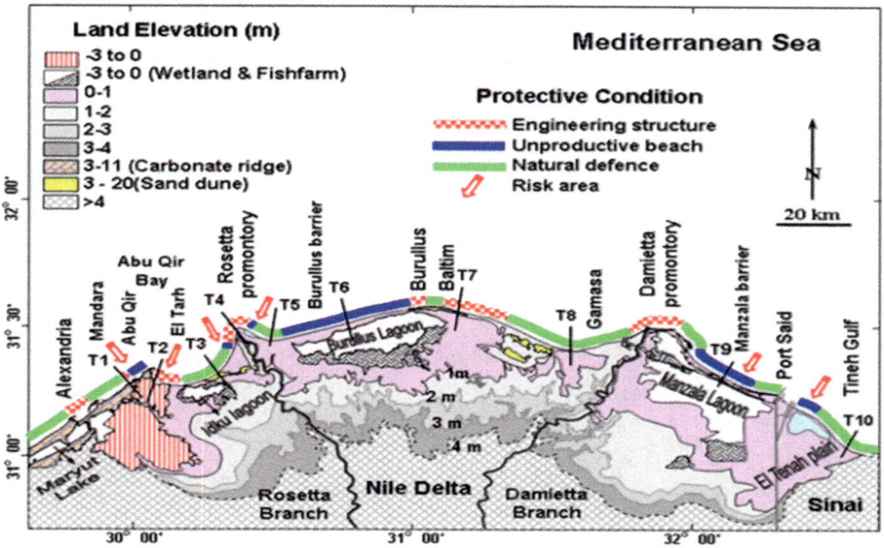

Fig. 7.9 Engineering defences along Nile Delta. (Source: Omran E. Frihy and Mahmoud Kh. El-Sayed, Springer)

Kiribati	Kiribati Adaptation Programme project (Kiribati seawalls)
Description	Kiribati is a small island country in the central Pacific Ocean that is very vulnerable to SLR. The project consists of three phases: Phase I, included the preparation for mainstreaming adaptation in line of the national planning; Phase II, consisted of the assessment and design adaptation of project framework; and Phase III, took in the planting 37,000 mangroves, constructed seawalls and the improvements in water supply.
Scope	Climate change adaptation
Timeline	2003–2016
Contractor	Government of Kiribati
Total cost	10 million US$
Partner	Least Developed Country Fund Japan Policy and Human Resources Development Fund The Global Facility for Disaster Reduction and Recovery
Aim	Improve water resource use and management Increase coastal resilience Strengthen capacity to manage SLR
Challenge	Negative impact of seawalls on ecosystems
Outcome	Integrated assessment database Water management result in save 271 cu.m/day Protect 1.18 km of coastline Increase coastal resilience
Sources	[27, 28]

Fig. 7.10 New seawall in Tarawa, Kiribati. (Source: Simon Donner)

Table 7.7 Weight evaluation criteria of M. Ali and Kiribati seawall projects

Active evaluation criteria	Mohamed Ali Seawall Project  Egypt	Kiribati's Adaptation Programme Project *(Kiribati seawalls)*  Kiribati
Relevance	1	1
Stakeholders' participation	0	1
Flexibility	0.5	1
Effectiveness	1	1
Sustainability	0	1
Equity	0.5	1
Efficiency	1	1
Total	4	7

Evaluation

Coastal protection aims at mitigating inundation and erosion due to sea level rise (SLR); hence both projects are very relevant to CCA. In Mohamed Ali Seawall Project, decision-making is limited to authorities; thus it scores 0 in Stakeholders' participation criterion. In contrast, the Kiribati project involves local communities in implementation through encouraging people to build seawall with their natural local materials to allow shoreline sediment. It also encourages and assists them to restore shoreline vegetation that reduces the negative effect of engineering measures and enhances ecosystem. This resulted in the project achieving the highest score in Stakeholders' participation evaluation criterion and Sustainability criterion as shown in Table 7.7. Engineering measures in Alexandria's seawall have negative effects on ecosystem near the shoreline; hence, it scores 0 in sSustainability criterion. The Alexandria seawall is expected to convert total affected area by 2025 from 701 Km^2 to 153 Km^2, whereas Kiribati project protects 1.18 km of coastline; thus both projects achieved the highest score in Effectiveness evaluation criterion.

Complementary programmes in Kiribati project restore the ecosystem and improve living condition of local community, granting Sustainability criterion and Flexibility criterion, unlike the M. Ali's project, which its benefits are reached under SLR only. Both projects are efficient because economic benefits exceed the budget costs.

According to the pervious evaluation, Kiribati seawall project is highly satisfactory in terms of the overall evaluation, but M. Ali's seawall is unsatisfactory due to inflexibility, non-sustainability and absence of stakeholders' involvement.

 Egypt	Participatory Development Programmes (PDP) in Urban Areas
Description	This is a collaborative programme between the Egyptian Ministry of State for Urban Renewal and Informal Settlements and Deutsche Gesellschaft für Internationale Zusammenarbeit (GIZ) GmbH (German Development Cooperation) The programmes focus on developing informal areas in Greater Cairo by fostering implementation of participatory policy tools, networking mechanisms among public, civil and private sectors to improve environmental conditions and achieve better service delivery in poor urban areas Since 2011 the programmes incorporated climate change adaption and urban resilience in urban development to promote resilient communities and urban areas
Scope	Climate change adaptation
Timeline	2004–2018
Contractor	Ministry of Planning, Egypt (MoPl)
Total cost	49 million EUR
Partner	Deutsche Gesellschaft für Internationale Zusammenarbeit, GIZ European Union, EU
Aim	Improve environmental conditions and better service delivery Implement sustainable solutions for informal area Foster climate change adaptation and urban resilience
Challenge	Poor swage and solid waste management Weak enforcement of building regulations and laws Unemployment
Outcome	Enhancing participatory approach in informal settlements Ensuring the sustainable development of informal areas Improving public amenities and services Enhancing socio-economic among local community Effective solid waste management (SWM) Maximizing job opportunities Improving public space and health care Raising efficiency of infrastructure
Sources	[29–31]

Fig. 7.11 Rooftop grading in Cairo. (Source: GIZ)

 Philippines	Strengthening Urban Resilience for Growth with Equity (SURGE) Project (Philippines)
Description	The project merges between sustainable development (SD) and adaptation to promote competitiveness, reduce vulnerability to climate change and enhance equity in formal settlements in Metro Manila, Cebu and Davao in the Philippines
Scope	Climate change adaptation and mitigation
Timeline	2015–2020
Contractor	Government of the Philippines
Total cost	47 million US$
Partner	US Agency for International Development (USAID)
Aim	Strengthening resilience through urban development Investing in low-emission local economic development through improved infrastructure and transport systems Expanding economic connectivity between urban and rural areas Improving live condition
Challenge	Flooding, sea level rise and climate change The gap between urban and rural areas Rapid economic activity
Outcome	Inclusive growth Pave away of less developed cities for growth. Resilient urban cities
Sources	[32, 33]

Fig. 7.12 Informal settlement and economic hub of Manila. (Source: Chamber of real estate and builders' associations)

Table 7.8 Weight evaluation criteria of PDP and growth with equity projects

Active evaluation criteria	Participatory Development Programmes (PDP) in Urban Areas  Egypt	Strengthening Urban Resilience for Growth with Equity  Philippines
Relevance	1	1
Stakeholders' participation	1	1
Flexibility	1	1
Effectiveness	1	1
Sustainability	1	1
Equity	1	1
Efficiency	1	1
Total	7	7

Evaluation

Informal settlements result in significant deterioration in quality of life with low access to infrastructure that would double the impact of CC. The two projects aim at mitigating this effect by following a participatory development approach through collaboration between the society and public administrations to promote sustainability, support sustainable developments goals (SDGs) and respect equity. Hence, both projects depend on lateral approach for stakeholders' engagement, to ensure the achievement of the project's objectives and planned direction.

The two projects are very relevant to CC adaptation, and positive effects were also reached without climate change risk (CCR), so the two projects get the highest score in every evaluation criteria as shown in Table 7.8, and the overall performance is highly satisfactory.

7.3.4 CCA Projects' Case Studies in Ecosystem Sector (Figs. 7.13 and 7.14)

 Egypt	Adaptation to Climate Change in the Nile Delta through Integrated Coastal Zone Management (ICZM PROJECT)
Description	The project seeks to integrate climate change adaptation into the coastal development by integrating coastal zone management into adaptation, strengthening the regulatory framework and building up institutional capacity. The project depends on environmentally friendly measures, as living shoreline approach by habitat restoration techniques and dune rehabilitation, to minimize coastal erosion and maintain ecosystems
Scope	Climate change adaptation
Timeline	June 2008 to June 2014
Contractor	Egyptian Environmental Affairs Agency (EEAA)
Total cost	16 million US$
Partner	UNDP Egypt
Aim	Implement innovative and environmentally friendly 'Living Shoreline' measures Monitoring and assessment adaptation framework
Challenge	Lack of baseline information on coastal processes Lack complementary Shore Management Framework (SMF)
Outcome	Sand dune restoration and management Beach drainage systems and lake bank stabilization Living shorelines Coastal development legislation and regulation focusing on ICZM
Sources	[34, 35]

Fig. 7.13 Wetland as part of the living shoreline approach. (Source: sustainable seas)

 India	India Integrated Coastal Management Programmes
Description	Integrated coastal zone management (ICZM) project is a process to integrate all aspects of sustainability in the management of the coastal zone, which aims at creating a balance between multiple human activities and conserving coastal ecosystem to build resilience to climate change
Scope	Climate change adaptation
Timeline	2011–2017
Contractor	Ministry of Environment and Forests
Total cost	285 million US$
Partner	International Development Association (IDA)
Aim	Vulnerability assessment to climate change Biodiversity conservation and promote better and sustainable livelihood
Challenge	Degradation of coastal ecosystem Conflict due to rural growth and rapid economic development
Outcome	Capacity building of scientific institutes and community Integrate of ICZM plan with environmental planning Infrastructure development includes rural electrification, eco-tourism facilities and multipurpose cyclone shelters
Sources	[36, 37]

Fig. 7.14 Waterfront development in Digha, India. (Source: IESWM, Dept. of Environment)

Table 7.9 Weight evaluation criteria of Egyptian and India ICZM projects

Active evaluation criteria	ICZM project – Egypt 	ICZM programme – India 
Relevance	1	1
Stakeholders' participation	0	1
Flexibility	1	1
Effectiveness	1	1
Sustainability	1	1
Equity	0.5	1
Efficiency	1	1
Total	5.5	7

Evaluation

Mitigating negative effects of SLR such as storm surge, flood and cyclone is one of the priorities of integrated coastal management that is very relevant to CCA. Both projects are planned to involve stakeholders in planning and monitoring, but only India's project succeeds in following the lateral approach, while Egypt's project limits decision-making to public administration and scored lowest in the stakeholders' participation criterion and equity as shown in Table 7.9.

Positive effects of the ICZM project were also reached without climate change risks (CCR) due to the fact that the project promoted sustainable benefits on environment, social and economic condition of community. Both projects are cost-efficient and succeed in reaching their objectives through increasing resilience, reducing vulnerability to SLR and building up capacity.

The overall performance of the Indian project is highly satisfactory, while that of Egypt's project is moderately satisfactory due to weak stakeholders' participation in adaptation process.

7.3.5 CCA Projects' Case Studies in Awareness Raising Sector (Figs. 7.15, 7.16, 7.17, and 7.18)

 Egypt	Coping with Climate Change Friedrich-Ebert-Stiftung (FES)
Description	The project is developed by Friedrich-Ebert-Stiftung (FES), and it started in 2013 It aimed at carrying out studies to reduce the negative effects of climate change, guiding citizens and raising their awareness about the issue of CCR in Egypt governorates and the role of the citizen in adaptation to climate change. It also supports key actors in addressing climate change. Programme covers sector as health care, waste management, fishers and awareness raising
Scope	Climate change adaptation
Timeline	2013–2018
Contractor	Egyptian Environmental Affairs Agency (EEAA)
Total cost	N/A
Partner	Federation of Environmental Associations (FEA) Ministry of Environment (EEAA)
Aim	Supporting the Ministry of Environment and its technical apparatus to integrate CC into their policies Involve civil society organizations in climate change adaptation framework
Challenge	Low awareness raising about CC Low CC data available
Outcome	Training staff of Egyptian Environmental Affairs Agency Awareness raising of the society through a prepared citizens' guidebook on CC. Twenty five awareness raising programmes were held for NGOs and planned to hold additional forty five programmes about studies and simplified recommendations to deal with CC
Sources	[38, 39]

Fig. 7.15 Citizen's guide for CCA measures. (Source: Friedrich-Ebert-Stiftung)

 Solomon Islands	Community Resilience to Climate and Disaster Risk in Solomon Islands Project
Description	The project was initiated to strengthen climate, disaster risk information and early warning systems to increase the capacity of rural communities and encourage community investments in climate change adaptation and disaster risk reduction
Scope	Climate change adaptation
Timeline	2014–2019
Contractor	Ministry of Environment and Conservation (MoEC)
Total cost	9 million US$
Partner	Global Environment Facility (GEF) Global Facility for Disaster Reduction and Recovery
Aim	Integrated CCA in national development Establish an early warning network Support adaptation investments
Challenge	Natural disaster due to location in the Pacific Ring of Fire and the tropical cyclone belt
Outcome	Establishing the foundations of a national risk information to improve risk management Risk analysis, design, advisory and supervision services for climate risk management plans
Sources	[40, 41]

Fig. 7.16 Community consultations in Solomon Islands. (Source: Secretariat of the Pacific Regional Environment Programme)

Table 7.10 Weight evaluation criteria of Friedrich and Solomon Islands projects

Active evaluation criteria	Coping with Climate Change Friedrich-Ebert-Stiftung  Egypt	Community Resilience to Climate and Disaster Risk in Solomon Islands Project  Solomon Islands
Relevance	1	1
Stakeholders' participation	1	1
Flexibility	1	1
Effectiveness	1	1
Sustainability	1	1
Equity	1	1
Efficiency	1	1
Total	7	7

Evaluation

The two projects were initiated to build up capacity of the rural communities and government authorities to address climate change and enhance sustainability.

Collaborative arrangement between government and stakeholders in decision-making was conducted through the lateral approach to assist in unifying efforts of the projects. Egypt's and Solomon Islands' projects are flexible enough to deal with uncertainty through the ability of self-organization during the phases of both projects. The two projects also succeed in raising stakeholders' awareness and providing training, advisory and supervision services.

Moreover, the projects gave attention to the most vulnerable groups. Egypt's and Solomon Islands' projects are also cost-efficient due to higher benefits gained in comparison to costs of implementation.

Finally, both projects scored highest in every evaluation criterion, and the overall performance is highly satisfactory as presented in Table 7.10.

 Egypt	Assessment and Strategy Development to Respond to the Impact of Rising Sea Level on Human Mobility in Egypt
Description	This project supports Egypt's efforts to identify and address human mobility posed by climate change on vulnerable populations It also supports the integration of migration and human security issues into policy development and planning and test awareness initiatives to help in responding to SLR induced mobility
Scope	Climate change adaptation
Timeline	December 2009 to January 2010
Contractor	Ministry of Manpower and Immigration Ministry of Environment Coastal Research Institute
Total cost	200,000 US$
Partner	International Organization for Migration (IOM)
Aim	Understand potential impacts of SLR on migration and human security Support integration of migration and human security into policy planning Enhance awareness raising among key stakeholders
Challenge	Located south area of Abu Qir Bay below sea level Populated urban areas and deterioration of socio-economic condition
Outcome	Mapping hazards, topography and population density of vulnerable areas through GIS (geographic information systems) Assessed and potential measures to mitigate the implications on human disaster identified Increased awareness of community and build up capacity of CSOs, local institutions and students
Sources	[42, 43]

Fig. 7.17 Policy recommendations and practical proposals report. (Source: Patrizio Fanti, Programme Officer)

 Mauritius	Looking beyond Life-Saving Assistance to Flood-Affected IDPs in Gambella Region (Mauritius)
Description	The project contributes towards building the capacity of Mauritius Government and communities to deal with the challenges of SLR. As well as, the project enhances national efforts to implement climate change measures and formalize new policy for mitigating environmental migration
Scope	Climate change adaptation
Timeline	January 2010 to January 2011
Contractor	International Organization for Migration (IOM)
Total cost	200 million US$
Partner	The Indian Ocean Commission (COI)
Aim	Awareness raising and capacity building of Mauritius' institutions Research on migration, environment and disaster risk reduction
Challenge	Sea level rise and tropical cyclones Employment and recession Lack of quantitative data
Outcome	Enhance job opportunities in the environmental sector (recycling, eco-tourism, renewable energies) to contribute to sustainable development Raise stakeholders' awareness to train and provide employment opportunities Compile information through field assessments
Sources	[44, 45]

Fig. 7.18 Severe floods in Mauritius that killed eleven persons in 2013. (Source: Disaster Report)

Table 7.11 Weight evaluation criteria of Mobility and Gambella projects

Active evaluation criteria	Assessment and Strategy Development to Respond to the Impact of RSL on Human Mobility  Egypt	Looking Beyond Life-Saving Assistance to Flood-Affected IDPs in Gambella Region  Mauritius
Relevance	1	1
Stakeholders' participation	1	1
Flexibility	1	1
Effectiveness	1	1
Sustainability	1	1
Equity	1	1
Efficiency	1	1
Total	7	7

Evaluation

The projects mitigate forced migration resulting from climate change (CC) to the minimum extent possible. They involve government authorities through conducted workshops and collaborative arrangement between government and stakeholders in making decisions. These two projects also define stakeholders' role in the adaptation and describe further steps that can be implemented by stakeholders.

The Mauritian's project succeeded in dealing with CC uncertainty through concentrating on sustainable development (SD) and enhancing capabilities of community to support recycling, eco-tourism and renewable energies, which in turn support SDGs, whereas the Egypt's project slightly neglected this aspect. But, nevertheless, both projects succeeded in meeting their set main objectives through assessing risk and awareness raising about CC impact on migration among all stakeholders.

Both projects involved NGOs and local communities in the decision-making process and did not limit the benefit to a specific group. The overall outcome is highly satisfactory as shown in Table 7.11 due to higher benefits in comparison to low costs of implementation.

7.3.6 *CCA Projects' Case Studies in Agriculture Sector (Figs. 7.19 and 7.20)*

Egypt	Building resilient food security systems to benefit the Southern Egypt region
Description	The project was initiated by a joint effort with international organizations to build resilience of the agriculture sector in Southern Egypt This was carried out through efficient water practices, banning unsustainable agricultural practices by shifting to organic farming and providing climate-friendly houses as well as efficient irrigation, livestock assets, financial assets and institutional strengthening and, finally, adaptation guidance
Scope	Climate change adaptation
Timeline	2005–2010
Contractor	Ministry of Agriculture (MoAgr), Egypt
Total cost	9 million US$
Partner	Spanish Government Swiss Fund International Development Research Centre (IDRC)
Aim	Enhance water-saving irrigation and other adaptation techniques Establishment of agroforestry GHG and plots with sub-surface irrigation Build institutional capacity through training of government technical staff Adaptation through technology development and transfer
Challenge	Uncertainty of CC projections
Outcome	Improve the adaptive capacity of the southern zone Use water-saving irrigation and other adaptation techniques Training government technical staff Establishment of climate change and adaptation online application
Sources	[46, 47]

Fig. 7.19 Consultative meeting in the project – planning phase. (Source: Magda Ghonem, WFP Consultant)

 Central African	Combating the Effects of Climate Change on Agricultural Production and Food Security in Central African Republic (CAR)
Description	The project was a part of the first national poverty reduction strategy and aimed at responding to the impacts of climate change on agricultural production and food security in CAR through setting plans for managing climate change risks in the agricultural sector and adopting water management and soil fertility, pasture and rangeland management
Scope	Climate change adaptation
Timeline	2011–2014
Contractor	CAR Ministry of Agriculture (MoAgr) CAR Ministry of Environment (MoEnv)
Total cost	5 million US$
Partner	United Nations Development Programmes (UNDP) GEF
Aim	Build capacity of farmers and institution Integration of CC in agriculture policy formation Strengthening climate risk management for enhancing food security and rural livelihoods
Challenge	Uncertainty of CC projections
Outcome	Develop policy, institutional and financial capacities for managing climate change risks in the agriculture sector Adapt agro-pastoral practice and technologies (water management and soil fertility, pasture and rangeland management) implemented in key vulnerable areas
Sources	[48, 49]

Fig. 7.20 Soil erosion in Central African Republic. (Source: Pinterest)

Table 7.12 Weight evaluation criteria of Egypt and CAR projects

Closed evaluation criteria	Building Resilient Food Security Systems to Benefit the Southern Egypt  Egypt	Combating the Effects of Climate Change on Agricultural Production and Food Security in CAR  Central African
Relevance	1	1
Stakeholders' participation	1	1
Flexibility	1	1
Effectiveness	1	1
Sustainability	1	1
Equity	1	1
Efficiency	1	1
Total	7	7

Evaluation

The two projects measure climate change impacts on the agriculture sector. Both projects were built on a cooperation agreement between farmers and authorities.

These two projects indicate that positive effects were reached even without climate change risk (CCR). This happened mainly due to the fact of enhancing framers' living conditions and agricultural productivity.

Both South Egypt's and CAR's projects were effective in increasing resilience to CC and supporting goals (1st and 7th of the MDGs), so the projects are cost-efficient. Both projects scored highest in every evaluation criterion as illustrated in Table 7.12; hence the overall performance is highly satisfactory.

7.4 Results

The Egyptian adaptation strategy consists of nine climate change adaptation projects (CCAPs) mainly concentrating on six sectors: (1) energy, (2) waste management, (3) urban, (4) ecosystem, (5) agriculture and (6) awareness raising. Approximately 70% of the adaptation projects share concern on three themes, energy, urban and awareness raising, whereas the rest are focusing on waste management, ecosystem and agriculture as shown in Fig. 7.21. Despite the fact that most of the vulnerable areas are being covered, the number of projects is modest in relation to the overall risks.

7.4.1 Evaluation of CCAP Case Studies in Energy Sector

Figure 7.22 illustrates the evaluation of CCAP's case studies in the energy sector. It indicates that two Egyptian CCA projects performed well, and therefore, both projects score high in the relevance and flexibility evaluation criteria. This was through the ability in coping with CC sequences under different scenarios as well as dealing with uncertainty of CC. These projects are cost-efficient compared to the costs of alternative ways of producing similar results. In contrast, there are some deficiencies regarding three criteria: (a) the Sustainability criterion, (b) Stakeholders' precipitation criterion, and (c) the Justice criterion. This was due to the lack of stakeholders' engagement in decision-making and the adverse effect on birds' migration routes although these side effects are common in this type of projects globally. Medium effectiveness of the projects is due to the slowdown in reaching the set targets. This is a mere result of the political situation after January 25, 2011.

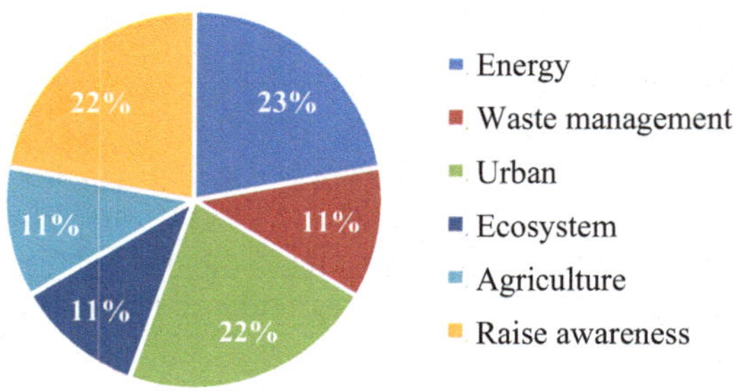

Fig. 7.21 Themes of CCA projects in Egypt. (Source: Disaster Report)

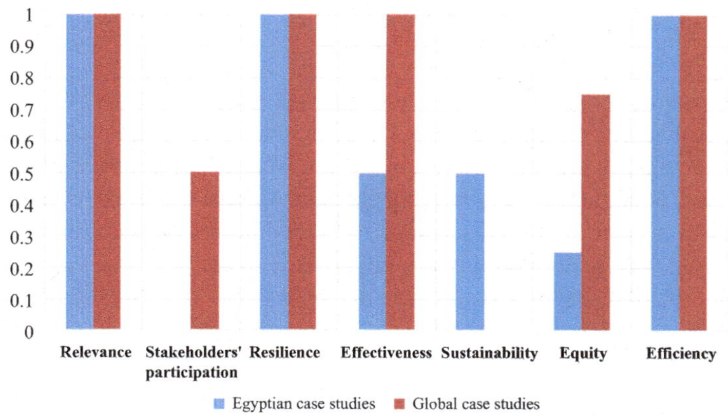

Fig. 7.22 Evaluation CCAP case studies in energy sector. (Source: Developed by authors)

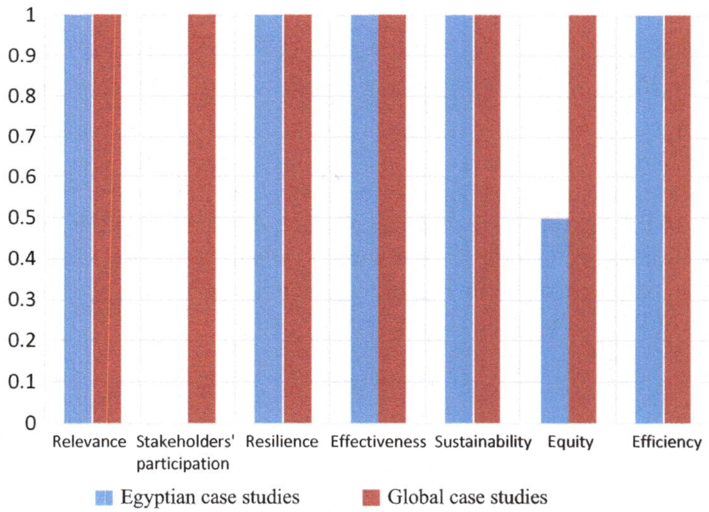

Fig. 7.23 Evaluation CCAP case studies in waste management sector. (Source: Developed by authors)

7.4.2 Evaluation of CCAP Case Studies in Waste Management

Climate change adaptation project (CCAP) in waste management sector is relevant to CC adaptation and mitigation and took into account sustainability through enhancing environmental, social and economic condition that provides higher benefits in comparison to project costs. The project achieves flexibility by not limiting benefits to a particular scenario. Limiting decision-making to local authorities led to reducing stakeholders' participation as shown in Fig. 7.23.

7.4.3 Evaluation of CCAP Case Studies in Urban Sector

The performance of CCAP in the urban sector is highly satisfactory in terms of three criteria (the Relevance, Effectiveness and Efficiency). This resulted from the success in reducing vulnerability to SLR in coastal areas and decreasing high temperature in informal settlement; higher benefits gained make the projects economically feasible. The projects achieved satisfactory level in terms of Flexibility and Equity criterion. This was by determining a balance between resilience and semi-resilience measures as well as drawing attention to the most vulnerable communities. In addition, Alexandria's seawall project depends on engineering measures that were considered to protect shorelines; and this led to negative effects on ecosystems in contrast to the participatory approach in the Kiribati project, which is an environmentally friendly. Moreover, the seawall project depends on a top-down approach whereas the second participatory project follows a lateral approach in Stakeholders' participation criterion. Therefore, the overall project performance in Sustainability and Stakeholders' participation evaluation criterion is unsatisfactory (Fig. 7.24).

7.4.4 Evaluation of CCAP Case Studies in Ecosystem Sector

Integrated coastal zone management projects are characterized as sustainable and capable of increasing the resilience of ecosystem environment to climate change under different scenarios. Egypt's projects succeed in achieving this, which makes the project in terms of criteria Relevant, Resilient, Effective, Sustainable and Efficient, but it fails in drawing attention to the most vulnerable people and involving them in decision-making as illustrated in Fig. 7.25.

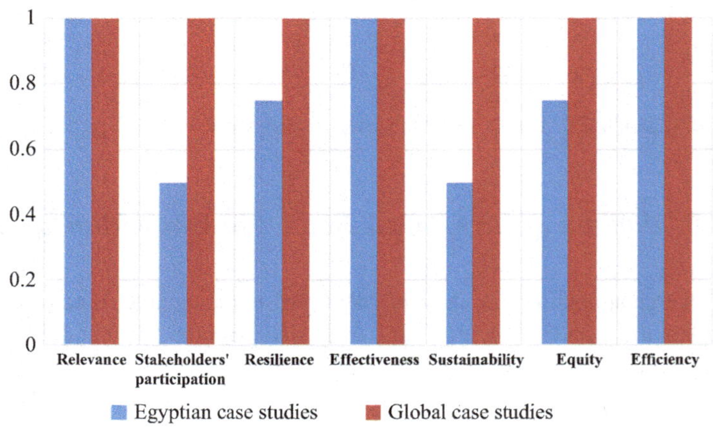

Fig. 7.24 Evaluation CCAP case studies in urban sector. (Source: Developed by authors)

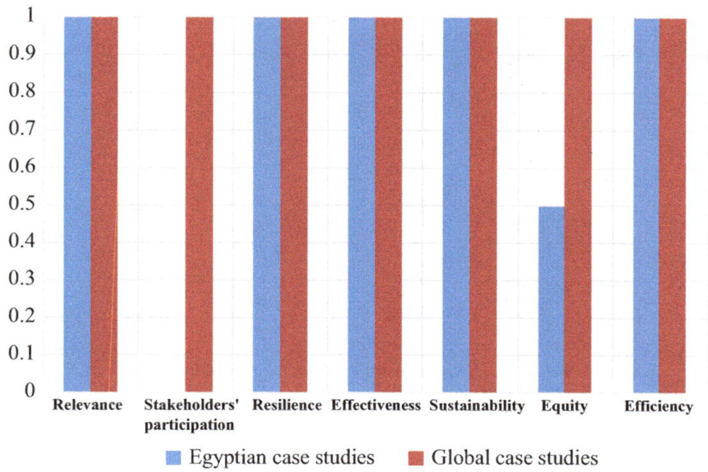

Fig. 7.25 Evaluation CCAP case studies in ecosystem sector. (Source: Developed by authors)

7.4.5 Evaluation of CCAP Case Studies in Awareness Raising Sector

The projects aim at building capacity of stakeholders to address CC and promote sustainability to deal with the uncertainty. The projects succeeded in awareness raising and building stakeholder's capabilities while considering equalities between societies and involving NGOs and local communities in decision-making which grants the project a level of highly satisfactory in criteria of Relevance, Stakeholders' participation, Flexibility, Effectiveness, Equity and Efficiency as well as highly satisfactory in Sustainability as depicted in Fig. 7.26. However, they need additional attention to mitigate projected migration in Northern Egypt and Delta due to SLR.

7.4.6 Evaluation of CCAP Case Studies in Agriculture Sector

Climate change adaptation projects (CCAPs) in the agriculture sector contain one project that mitigates climate change risks (CCR) in Southern Egypt farmlands through the cooperation between farmers and authorities to promote sustainable agricultural practices and provide climate-friendly houses for vulnerable people. The projects are effective in increasing resilience to CCR and support goals 1 and 7 of the MDGs, which make the projects cost-efficient. The overall evaluation outcome is highly satisfactory as depicted in Fig. 7.27.

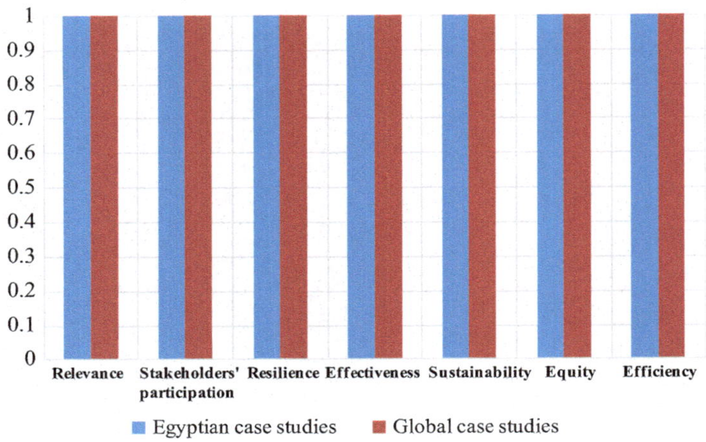

Fig. 7.26 Evaluation CCAP case studies in awareness raising. (Source: Developed by authors)

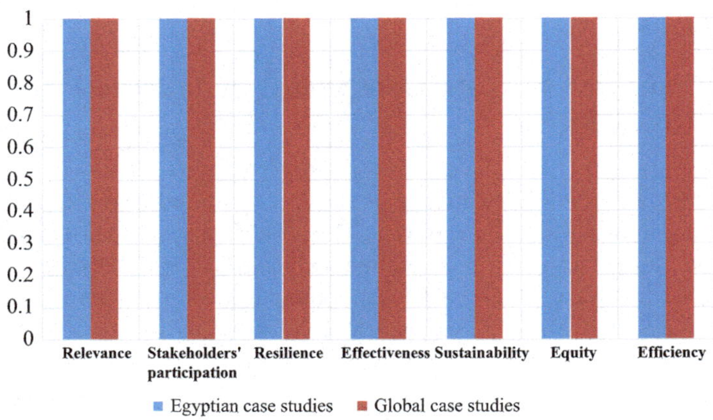

Fig. 7.27 Evaluation CCAP case studies in agriculture sector. (Source: Developed by authors)

7.5 Discussion

The overall evaluation of case studies in comparison to global CCAP is illustrated in Fig. 7.28. It indicates that the Egyptian projects met the performance of global projects in criteria of Relevance, Flexibility and Efficiency due to choosing resilient measure that succeeds in coping with adverse effects of climate change through reduced exposure, sensitivity and increased adaptive capacity under different scenarios. The projects are cost-efficient due to higher benefits gained which make them economically feasible. Egypt's CCAPs are slightly low in terms of

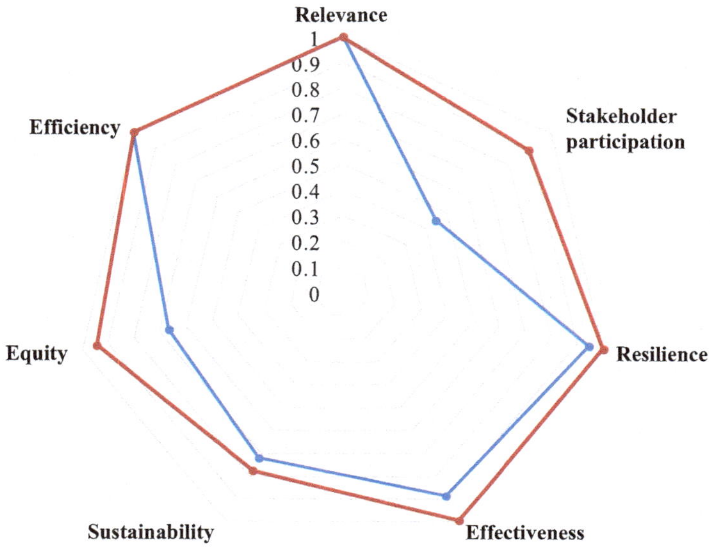

Fig. 7.28 Evaluation case studies of CCAP in all sectors. (Source: Developed by authors)

sustainability, despite them enhancing social and economic condition and drawing attention to the most vulnerable communities. However, projects need to stress on enhancing environmental condition. Moreover, projects are a little bit low in terms of effectiveness due to delays in wind power project timetable caused by political situation after January revolution, but authorities pledged to finish this project on schedule.

There is a wide gap in stakeholders and equity performance because most of the projects did not make sufficient effort to involve stakeholders and depended on top-down approach in most of projects' phases. The gap between Egyptian adaptation projects and global projects due to weak stakeholders' involvement led to weak performance in equity and stakeholders' participation. For the Mohamed Ali's sea-wall project and Egypt's wind power project, both are considered unsustainable due to two reasons: (a) negative impacts on ecosystem (though it is sustainable due to the clean energy generated), and (b) delays in wind power projects' timetable due to the political situation after January 2011. All resulted in the overall performance of Egypt's CCA project to be moderately satisfactory and do not meet the Performance Criterion of the global CCA case studies, which is highly satisfactory (Fig. 7.29).

The next section provides the recommendations for enhancing the performance of Egyptian CCA projects to reduce the gap, as a demonstration example that can be followed in other developing countries.

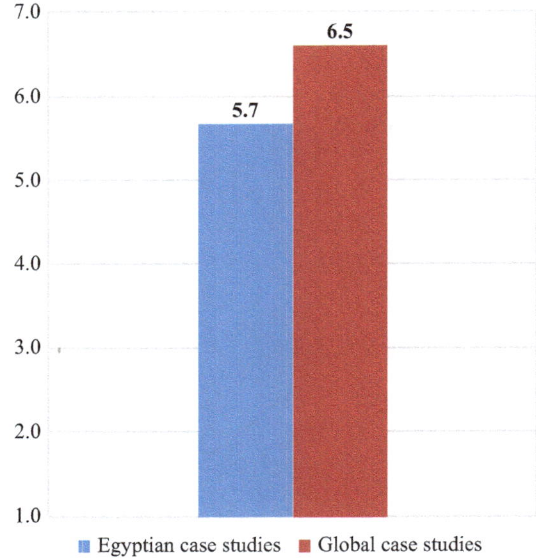

Fig. 7.29 Evaluation overall performance of CCAP case studies in all sectors. (Source: Developed by authors)

7.6 Recommendations for Enhancing Egypt's CCA Projects

The dilemma of inadequate stakeholders' involvement in Egyptian adaptation projects undermine the efforts of the project in achieving maximum benefit and led to weak performance in Equity and Stakeholders' precipitation criteria. Authorities should start promoting equity in climate change adaption projects and not just considering equalities between sectors and regions. In addition to giving attention to the most vulnerable groups, the authorities should also involve a broad range of stakeholders from different sectors in all stages of adaptation through involving them in planning, prioritization, monitoring and evaluation and enhance clarity of adaptation process and transparency of decision-making by providing clear and accurate information through participatory adaptation. Participatory approaches for stakeholders' engagement combine between policy-based approaches to support implementation in addition to ensuring society acceptance and science-based approaches by integrating stakeholders into the research process to acquire practical and local knowledge regarding correct adaptation process [50]. Figure 7.30 shows Cairo Climate Talks, one of the awareness raising platforms dedicated to exchanging experiences and fostering cooperation between experts and civil society in field of climate change. According to the United Nations Framework Convention on Climate Change (UNFCCC), effective stakeholders' engagement process should take into account certain actions: define challenges and objective of adaptation, clarify reflection of engagement process on official decision-making, identify stakeholders in

Fig. 7.30 Podium discussion about awareness raising and fostering cooperation between experts and civil society in Cairo. (Source: The Cairo Climate Talks)

open and transparent manner, ensure access to information and build trust and develop capacities of stakeholders [51].

The poor performance of the Egyptian projects in Sustainability criterion is due to the negative impacts of Egypt's wind power project and Mohamed Ali seawall on the environment. Wind power project has adverse effect on birds due to isolating migration routes and causing collisions of birds with turbines. Mohamed Ali seawall has negative impact on shoreline ecosystem; scientific studies documented a loss of ecosystem services, loss of habitat and reduction in biodiversity as result of engineering measures. As a result, promoting sustainability in Egyptian adaptation projects should be done through re-evaluating the environmental impact of Egypt's wind power and the adverse effect on birds through initiating parallel programmes to reduce the consequences of isolating migration routes and building up capacity of scientific research to reduce information gap about consequences of the collision of birds with turbines.

The Federal Aviation Administration in the United States recommends installing aviation obstruction light on wind turbine as illustrated in Fig. 7.31 to warn birds in dark nights; it also recommends shutting down or curtailing blade rotation during bird migration waves and coating blades with ultraviolet paint to increase turbine visibility and reduce collision with birds [52]. Mohamed Ali Seawall Project also has negative impact on intertidal ecosystems and bird breeding sites, so initiating a programme to monitor and assess the damage and prepare seawall management plan to recover ecosystem is a must. Moreover, doing some engineering modifica-

Fig. 7.31 Wind turbine obstruction light. (Source: Ucakikazlambasi)

Fig. 7.32 Ecological coastal armouring units contain holes to provide habitats for organisms. (Source: DECCW and Sella Ido)

tions on future seawall structure such as crevices, holes and slopes to encourage marine growth through working as a habitat is encouraged [53].

Figure 7.32 presents ecological coastal armouring units that contain holes to provide habitats for organisms. Also, it was noticed that the lack of performance of sustainability in the Integrated Solar Combined Cycle Power Project in Morocco is basically due to involuntary re-settlement. According to World Bank, this has a negative effect on the 41% of active bank-financed project which is physically displaced, and facing impacts on livelihoods. In this regard, the World Bank's preference is to avoid relocation as much as possible or, in the case of inevitability, impose re-settlement action plans [54]. The problem of the delay in the wind power project timetable was due to political situation after January 2011. However, this was solved through paving the way for private sector to invest in renewable energy projects in 2016 and by removing obstacles that hinder the implementation of feed-in tariff projects [54].

7.7 Conclusion

Climate change has enormous negative risks according to Global Humanitarian Forum. Five hundred million people, most of them in developing countries, are at extreme risk. Therefore, governments and civil society in developing countries attempt to address climate change through adaptation projects and programmes, as well as continuous monitoring and evaluating projects to ensure that they are capable of meeting the targeted objective.

Comparing between Egyptian CCA projects and other case studies from developing countries, assessing the gaps and tracking successes and failures in order to learn and increase knowledge indicate the following main finding points. First, the performance of Egyptian CCA projects is highly satisfactory in terms of relevance, flexibility, effectiveness and efficiency due to choosing resilient measures and succeeding in coping with the adverse effects of CC; as well as, the projects are also cost-efficient because high benefits gained make projects economically feasible. Second, the projects are moderately satisfactory in terms of sustainability and equity and unsatisfactory in terms of stakeholders' participation. As a result, projects need to stress on enhancing the environmental condition and make sufficient effort to involve stakeholders in decision-making. Based on all of the above, the overall performance of the Egyptian CCA projects is satisfactory while that of global CCA case studies is highly satisfactory, so authorities should reduce the gap to reach the best CCA project performance. This can be done by enhancing the performance of stakeholders' participation through involving a broad range of stakeholders in planning, prioritization, monitoring and evaluation and enhancing clarity of adaptation process and transparency of decision-making to support implementation in addition to ensuring society acceptance.

The performance of the Egyptian projects in sustainability can be enhanced by initiating parallel programmes to reduce the consequences of isolating migration routes by wind farms through installing aviation obstruction light on wind turbine or curtailing blade rotation during bird migration waves. The negative impact of Mohamed Ali seawall on intertidal ecosystems can be mitigated through recovering ecosystem and doing some engineering modifications on future seawall structure such as crevices, holes and slopes to encourage marine growth. The delay in wind power projects' timetable can be solved through paving a way for a private sector to invest in renewable energy by solving obstacles that hinder the implementation of feed-in tariff projects.

Nevertheless, as of 2015, the Government of Egypt has taken bold steps towards attaining sustainable development goals (SDGs) and climate change adaption (CCA), which were integral parts of Egypt's vision 2021 and Egypt's Sustainable Development Strategy 2030. In addition, on July 1, 2018 Egypt presented the National Voluntary Review report at the United Nations. In this report, climate change adaptation and implementation of Sustainable Development Goals (SDGs) were reported with a special focus on SDG 13 – Climate Action and its related indicators. More climate change adaption projects (CCAPs) were launched in 2016, 2017 and 2018.

References

1. "Human Impact Report: Climate Change- The Anatomy of a Silent Crisis", Global Humanitarian Forum, 2009.
2. "Assessing the Costs and Benefits of Adaptation Options", United Nations Climate Change Secretariat, Bonn, Germany, 2011.
3. Merylyn Hedger et al, "Evaluation of Adaptation to Climate Change from a Development Perspective", GEF Evaluation Office, 2008.
4. Rob D. van den Berg and Osvaldo N. Feinstein, Evaluating Climate Change and Development, Transaction Publishers, 2011.
5. Jo-Ellen Parry and Susan Taylor, "Mainstreaming Adaptation to Climate Change into National Policy: An overview for adaptation practitioners", International Institute for Sustainable Development, August 2012.
6. Sniffer's technical advisory group, "Climate change adaptation-related indicators", Sniffer, Scotland, UK, July 2012.
7. Jouni Paavola and W. Neil Adger, "Justice and Adaptation to Climate Change", Tyndall Centre for Climate Change Research Working Paper 23, Norwich, UK, October 2002.
8. LDC Expert Group, "National Adaptation Plans Technical guidelines for the national adaptation plan process", United Nations Climate Change Secretariat, Bonn, Germany, December 2012.
9. J. R. A. Butlera et al, "Integrating Top-Down and Bottom-Up Adaptation Planning to Build Adaptive Capacity: A Structured Learning Approach", *Coastal Management,* vol. 43, no. 4, pp. 346–364, 2015.
10. "Rating the Outcome of Projects with Formally Revised Objectives: Joint Guidelines for Use in ICRs and in OED Assessments", The World bank, 2005.
11. "Orascom construction official website", [Online]. Available at: www.orascom.com/our-capabilities/case-studies/kuraymat-solar-power-field/. (Accessed: May 5, 2016).
12. "Kureimat Solar Thermal Hybrid Project", World Bank official website, [Online]. Available at: www.worldbank.org/projects/P050567/kureimat-solar-thermal-hybrid-project?lang=en. (Accessed: May 5, 2016).
13. "Egypt and Morocco : Concentrated Solar Power", World Bank official website, [Online]. Available at: http://web.worldbank.org/WBSITE/EXTERNAL/COUNTRIES/MENAEXT/0,,contentMDK:22866325~menuPK:3949143~pagePK:146736~piPK:226340~theSitePK:256299,00.html. (Accessed: May 5, 2016).
14. Omar Behara et al, "A review of integrated solar combined cycle system (ISCCS) with a parabolic trough technology", *ScienceDirect,* vol. 39, pp. 223–250, July 2014.
15. "Global Environment Facility Trust Fund Grant Agreement", Office National De L'electricite, April 2007.
16. "Project Appraisal Document on a Proposed Grant from The Global Environment Facility Trust Fund in the Amount of US$43.2 Million," The World Bank Sustainable Development Department Middle East And North Africa (MNA), February 2007.
17. "Samallout/Suez Gulf/Jabal El-Zayt 500 kV Electrical Interconnection Project Environmental and Social Assessment and Management Plan", Egyptian Electricity Transmission Company, March 2010.
18. "Implementation Status & Results Egypt, Arab Republic of Egypt - Wind Power Development Project (P113416)", World Bank, 2014.
19. "Devex, Wind Umbrella Project In Mexico", [Online]. Available at: www.devex.com/projects/pipelines/5228. (Accessed: June 21, 2017).
20. World Bank. 2006. Mexico - Wind Umbrella (La Venta ii) Carbon Finance Project (English). Washington, DC: World Bank. Available at: http://documents.worldbank.org/curated/en/462021468123559294/Mexico-Wind-Umbrella-La-Venta-ii-Carbon-Finance-Project. (Accessed: May 19, 2016).

21. "Project design document onyx Alexandria landfill gas capture and flaring Project", United Nations Framework Convention on Climate Change, April 2006.
22. "Environmental impact assessment al hammam landfill project", Integral Consult, August 2005.
23. "ONYX solid Waste Alexandria - Carbon", The World Bank, [Online]. Available at: www.worldbank.org/projects/P098737/onyx-solid-waste-alexandria-carbon?lang=en&tab=results. (Accessed: May 19, 2016).
24. World Bank. 2008. Brazil - Second Novagerar Carbon Finance and Solid Waste Management Project (English). Washington, DC: World Bank. Available at: http://documents.worldbank.org/curated/en/669241468231893282/Brazil-Second-Novagerar-Carbon-Finance-and-Solid-Waste-Management-Project. (Accessed: May 8, 2016).
25. Mostafa Kamal Tolba et al, "Egypt second national communication under the United Nations framework convention on climate change,", Egyptian Environmental Affairs Agency, May 2010.
26. Jeffress Williams and Nabil Ismail, "Climate Change, Coastal Vulnerability and the Need for Adaptation Alternatives: Planning and Design Examples from Egypt and the USA", *Journal of Marine Science and Engineering,* vol. 3, no. 3, pp. 591–606, July 2015.
27. "Kiribati Adaptation Program official website", Government Of Kiribati, [Online]. Available at: www.climate.gov.ki/category/action/adaptation/kiribati-adaptation-program/. (Accessed: June 14, 2016).
28. "Kiribati Adaptation Phase III, Implementation Status & Results Report", March 2016, The World Bank.
29. "Participatory Needs Assessment In Informal Areas – Cairo Governorate", EcoConServ Environmental Solutions, August 2015.
30. Center for Development Service, "Participatory Needs Assessment in Informal Areas", Deutsche Gesellschaft für Internationale Zusammenarbeit (GIZ)GmbH, May 2013.
31. "About", GIZEG, [Online]. Available at: http://egypt-urban.net/about/. (Accessed: January 16, 2017).
32. "Strengthening Urban Resilience for Growth with Equity (SURGE) Project", USAID, [Online]. Available at: www.usaid.gov/philippines/economic-growth-and-trade/strengthening-urban-resilience-growth-equity-surge-project. (Accessed: June 27, 2016).
33. "Strengthening Urban Resilience for Growth with Equity (USAID/SURGE)", International City/County Management Association, [Online]. Available at: http://icma.org/en/international/projects/directory/Project/1074/Strengthening_Urban_Resilience_for_Growth_with_Equity_USAIDSURGE. (Accessed: June 21, 2016).
34. "UNDP Egypt", in *Inception workshop adaptation of the Nile delta to climatic changes and sea level rise through ICZM project*, 5–6 December 2010.
35. "Adaptation to Climate Change in the Nile Delta through ICZM Project", Sustainable Seas Ltd.
36. "ICZM Project", government of Odisha official website, [Online]. Available at: http://www.iczmpodisha.org/. (Accessed: June 28, 2016).
37. ICZM Project official website, [Online]. Available at: http://www.iczmpwb.org/. (Accessed: June 29, 2016).
38. Ahmed El Khouly and Ezzat Abdel Hamid, "Citizen Guide: The negative Impact on Climate Change", 2014.
39. "Coping with Climate Change report", Friedrich-Ebert-Stiftung, 2013.
40. Jordy, Denis Jean-Jacques, "Community Resilience to Climate And Disaster Risk In Solomon Islands Project: Implementation Status Results Report", World Bank Group, Washington, D.C., 2015.
41. "Environmental and social management framework. Solomon Islands", The World Bank, Solomon Islands, 2013.
42. Patrizio Fanti, "IOM approach to environmental induced Migration and Abu Qir Project", International Organisation for Migration, June 2015.

43. Angela Santucci, "International Organization for Migration (IOM)", in *Migration, Climate Change and the Environment Egypt Region project*, Cairo, 20–21 September 2010.
44. "Migration, climate change and the environment", International Organization for Migration, Geneva, Switzerland.
45. Alexandre Magnan, "IOM workshop "Climate change, environmental degradation and migration"", in *change environmental changes and migration in the republic of Mauritius*, Geneva, 2011.
46. "Project Proposal", Adaptation Fund Board Secretariat, 2012.
47. Magda Ghonem, " Mid-Term Evaluation-Building Resilient Food Security Systems to Benefit the Southern Egypt Region project", September 2015.
48. "CAR Project Identification Form", UNDP, November 2010.
49. Alec Crawford, Hilary Hove and Jo-Ellen Parry, "Review of Current and Planned Adaptation Action: Middle Africa Central African Republic", 2011.
50. Maja Rotter et al, "Stakeholder Participation in Adaptation of Climate Change", UMWELTBUNDESAMT, Roßlau, Germany, July 2013.
51. Anthony Nyong and Yvette Aguilar, "Engaging Stakeholders in the Adaptation Process", unfccc.
52. The American Wind Wildlife Institute, "Wind Turbine Interactions with Wildlife and their Habitats", The American Wind Wildlife Institute, June 2016.
53. "Guidelines for assessing the impacts of seawalls", Department of Environment, Climate Change and Water NSW, Sydney, 2012.
54. David Freestone, "The World Bank and Sustainable Development: Legal Essays", Martinus Nijhoff, Netherland, 2013.

Chapter 8
Climate Change Adaptation Scenarios in Developing Countries

Abstract This chapter presents the assessment of climate change (CC) impacts and scenarios as well as the simulation measures and tests of the applicability in urban areas in order to anticipate and predict climate change impacts and scenarios. It also illustrates the selected case study representing a housing project in an urban area in a developing country that is characterised as a hot-arid climate. The study simulates the urban microclimate in terms of the urban metaphor and setting from baseline year 2016 and compares it to that in 2080 in order to evaluate CC impact. In addition, the assessment tests six measures, including cooling passive design configurations, buildings' elevation, buildings' envelope, vegetation, water features and orientation and high albedo, to identify their impact on the adaptation scenarios to heat stress and comfort. The CC weather scenarios for 2080 were generated using CCWorldWeatherGen, which is based on the IPCC Third Assessment Report. This chapter also looks at the role of environmental design in CC adaptation. Since urban development is part of climate change adaptation action, it is imperative to investigate if such action was implemented on new projects and if forecast weather is meeting the CC adaptation measures.

8.1 Introduction

Urban areas in developing countries such as those located in the Middle East and North Africa (MENA) region are facing real challenges in adapting to climate change risks and in fulfilling thermal comfort in open urban areas as well. This colossal challenge is taking place, while global warming is rapidly increasing all over the world [1]. Meteoritic data indicate a rise in hot summer days [2] will result in a rapid increase in heat stress shocks and deaths' cases. These risks are not only limited to urban areas and public health, but also reflected on the energy sector through surging in the pressure of national electricity grids due to high demands in energy consumption for cooling and heating. Taking into account that most of these countries, which are located in MENA region, highly depends on fossil fuels for generating electricity, therefore greenhouse gas (GHG) emissions, mainly CO_2 will be released [3]. Hence, mitigation and adaptation action plans should be adopted to

M. M. Aboulnaga et al., *Urban Climate Change Adaptation in Developing Countries*, https://doi.org/10.1007/978-3-030-05405-2_8

curb such emissions. Although all of these risks seem complicated and interrelated, all problems indicate climate change (CC) is manifesting, however. Under these circumstances, it is necessary for architects and urban planners to simulate projected climate change scenarios and adapt to its outcomes. It is imperative to state that technology and simulation programmes, such as ENVI-met, play a major role in assisting planners and designers to develop resilient and sustainable cities that can withstand climate change severe events, besides, predict future climate change scenarios to ensure thermal comfort is achievable for better human well-being. Within the chapter, the role of buildings' retrofitting and cities' upgrading in coping with global warming is tested to assist in dwindling the information gap and achieving a deeper understanding of challenges and opportunities that cities face.

8.2 Adaptation Scenarios

This chapter investigates the applicability of adaptation strategies in adjusting to climate change. It also examines thermal comfort in urban areas of developing countries, specifically in a housing project located in Mokattam district, in the city of Cairo, Egypt. First, the scenario includes simulation of the long-term impact of climate change on Cairo urban areas by 2080, comparing it with the baseline year 2016; this is mainly to assess the impact of climate change. Second, the scenario tests the role of passive design configurations in mitigating the effect of warming in cities and climate change impact in order to identify possible countermeasures. Six proposed passive design configurations, for urban modifications, were tested to identify their role in adapting to heat stress in these urban areas.

These six strategies exploited in the assessment and simulations are illustrated in Fig. 8.1, as follows:

- *Buildings' Elevation*: elevates all buildings, in this case study; buildings were raised 3 m above the ground level, which represents one floor according to site regulations.
- *Buildings' Envelop*: enhances the thermal performance of exterior building walls and roofs to be compatible with international energy efficiency codes.
- *Vegetation*: increases vegetation in all areas of the site, including the empty sandy areas, the building's surroundings and between buildings.
- *Water Features*: adds water features in urban areas (hot arid areas), which were functionally acceptable such as in sandy areas around the city, gardens in the site and areas between buildings.

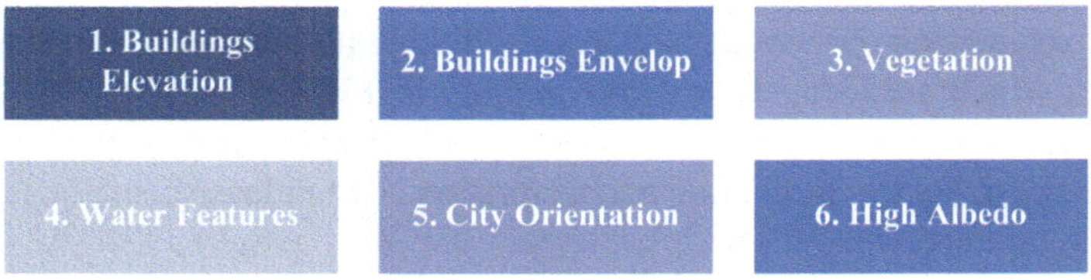

Fig. 8.1 Models simulated in research by ENVI-met. (Source: Developed by authors)

- *City Orientation*: orientates the urban settlement towards the main wind direction to catch prevailing winds to improve cooling when air temperature is below skin temperature 33 °C. This solution is not compatible with the case study, due to the fact that it already exists, but has been tested for evaluation, and the results can support in the future planning of cities.
- *High Albedo*: selects high albedo materials for buildings' facades, roofs and street pavements.

The simulations were run for each case on a typical summer day, July 1, 2017, for 14 hours from 05:00 to 19:00. Thematic maps and graphs were extracted at 08:00, 13:00 and 17:00 for the four climatic parameters that have the most influence on thermal comfort: dry bulb temperature (DBT), wind speed (V), relative humidity (RH) and mean radiant temperature (MRT). The predicted mean vote (PMV) is exploited to express the human perception of thermal comfort. Six proposal results were finally compared with base case to identify the effect of each of the passive design techniques on the thermal comfort by 2080.

8.3 Simulation Tools and Scenarios

Understanding of the simulation tools and skills in the field of environmental design to acquire the knowledge needed to perform climate change scenarios. This section presents some of the tools that were used in simulating the adaptation scenarios:

- *ENVI-met 3.1*: ENVI-met is a three-dimensional microclimate model designed to simulate the surface-plant-air interactions in urban environment with a typical resolution down to 0.5 m in space and 1–5 second(s) in time through using fundamental laws of fluid and thermodynamics. The programme is validated by many scientific researches.
- *LEONARDO 2014*: analyses and visualizes ENVI-met model results and creates 2-D and 3-D maps.
- *Energy Plus Website*: a weather data used in simulation was taken from Cairo International Airport weather station, which was provided by Energy Plus programme, validated by the US Department of Energy.
- *CCWorldWeatherGen*: generates climate change weather scenarios for 2020, 2050 and 2080 for worldwide locations to use in building performance simulation programmes. It is based on Intergovernmental Panel on Climate Change (IPCC) Third Assessment Report.
- *Climate calculator*: a software that calculates water content (g/kg) through input temperature, relative humidity and other default factors that are essential in ENVI-met input data.
- *PMV 2008 Version 1.0*: thermal comfort calculator software, which calculates predicted mean vote (PMV) – a scale used to predict thermal comfort. It also calculates the Predicted Percentage Dissatisfied (PPD), which predicts the percentage of occupants that will be dissatisfied with the thermal conditions.

8.4 Simulation Data Input

The simulation is carried out on a housing project 'Asmarat' in Cairo, Egypt, as illustrated in Fig. 8.2. The housing project encompasses 425 residential buildings, each consists six storeys in the first and second phases and ten storeys in the third phase. All these buildings were constructed with concrete and single layer walls with plasters. Streets and parking lots were paved with black asphalt, and sidewalks and shoulders' pavements were covered with sands, cements and interlocks (Fig. 8.3). Also, there is a lack of vegetation. It is selected as a case study because it is one of the latest governmental projects that aimed to eliminate informal housing and improve lives of their inhabitants. Therefore, it is important to align new projects with climate change adaptation policy by adapting with global warming as this is a part of the governmental vision of sustainable development strategy 'Egypt 2030'. Also, the site represents a model for high-rise buildings which can be found in many Egyptian cities, which enable the author to evaluate climate change impact on urban areas that are located in arid climate zones and evaluate the role of cooling passive design configurations in adaptation.

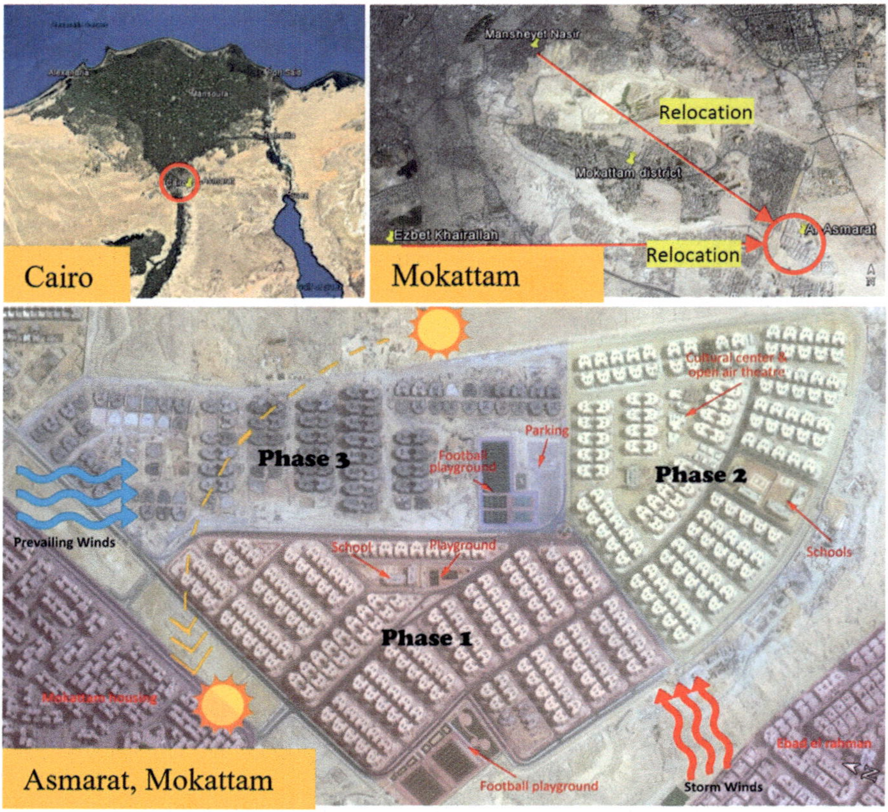

Fig. 8.2 Case study location. (Source: Google Earth)

Fig. 8.3 Buildings of the housing project – 'Asmarat'. (Source: alarabyanews.com/212494 (Accessed: November 26, 2017))

8.5 Data Input

The project layout was drawn by ENVI-met 3.1 software from the satellite images and site survey, and then the site location was defined on Cairo's longitude and latitude where the model area is on a $243x*152y*26z$ grid; size of grid cell is 4*4*3 in meters. Starting with the baseline case, the urban form was restructured including soils and pavements; textures in addition to green areas were specified. Twenty-two receptors were taken into consideration and added to monitor atmospheric parameters as presented in Fig. 8.4. The receptors are distributed to cover all model areas for more accurate results. The same process was repeated in all models. Tables 8.1 and 8.2 show the data input for weather and buildings for the following models (Fig. 8.5).

8.6 Scenario Description

- *As-built_2016*: the model represents as-built of the housing project 'Asmarat' under current weather situation. The EPW weather file used in the simulation was Cairo 2003 file for Cairo International Airport weather station, which is 13 km away from the simulated project.
- *As-built_2080*: the model represents as-built of the housing project 'Asmarat' under weather scenario for 2080 that projects an increase in surface temperature by approximately 4 °C based on Intergovernmental Panel on Climate Change (IPCC) Third Assessment Report. The CCWorldWeatherGen software was also used to generate 2080 weather scenarios.
- *Buildings' Elevation*: the model elevates all building height about 3.00 m above the ground, which represents one floor through clearing any wall on ground floor and leaving building columns only.
- *Buildings' Envelop*: the model that enhances the buildings' envelops was based on international energy efficiency codes to enhance thermal performance of exterior building walls and roofs, selecting materials with lower U-value (Wall = 0.57 W/m^{2}K, Roofs = 0.31 W/m^{2}K) for better insulation.
- *Vegetation*: the model increases vegetation in all areas of the site, adds grass on empty sandy areas surrounding the city and in yards and, also, adds ficus trees,

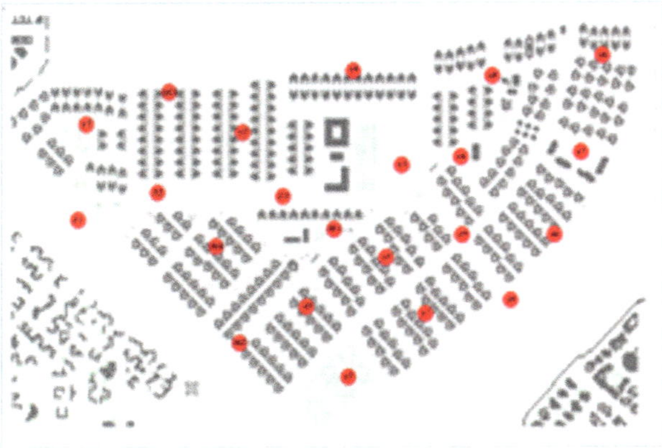

Fig. 8.4 Twenty-two receptors distribution in the site. (Source: Developed by authors)

Table 8.1 Input weather data needed for simulation

Parameter	Scenario 2016	Scenario 2080
Wind speed in 10 m ab. ground [m/s]	2.6	2.4
Wind direction (0:N, 90:E,180:S, and 270:W	330	330
Initial temperature atmosphere [K]	301.45	306.25
Specific humidity in 2500 m [g water/kg air]	9.42	11.48
Relative humidity in 2.00 m [%]	39	36

Table 8.2 Input building properties needed for simulation

Parameter	Simulation scenarios			
	Scenario 2016 Scenario 2080	Buildings' elevation Buildings' envelop	Vegetation Water features City orientation	High albedo
Inside temperature [K]	300.15	299.15	300.15	300.15
Heat transmission walls [W/ m^2K]	1.7	0.57	1.70	1.70
Heat transmission roofs [W/ m^2K]	2.0	0.31	2.00	2.00
Albedo walls	0.30	0.30	0.30	0.5
Albedo roofs	0.20	0.55	0.20	0.5

4.00 m dense, in streets and open areas for shading. Yellow poinciana trees, 20.00 m dense, were added surrounding the city to redirect the wind in addition to acting as a buffer zone to protect the city from sandy storms.

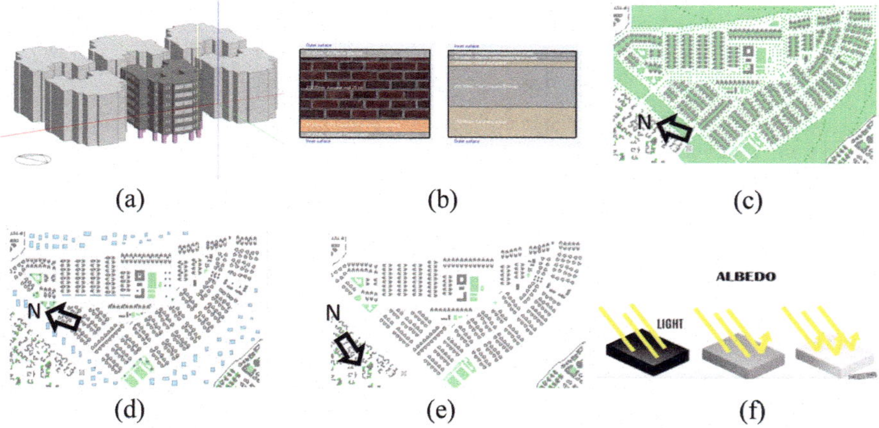

Fig. 8.5 Illustration of the adaptation strategies modelling. (**a**) Buildings elevation. (**b**) Buildings envelop. (**c**) Vegetation. (**d**) Water feature. (**e**) City orientation. (**f**) High Albedo impact. (Source: Developed by authors)

- *Water Features*: the model adds water features in empty sandy areas surrounding the city, building yards and green areas, which were functionally acceptable for adding a water feature.
- *City Orientation*: the model arranges the street canyon to be oriented towards the main wind direction, meaning that the street becomes in line with the prevailing wind direction which gives low resistance to air movement. This solution may not be compatible with the case study due to the fact that it already exists, but has been tested for evaluation to assist in the future panning of similar projects.
- *High Albedo*: the model selects high albedo materials for buildings' facades, roofs and street pavements. This is achieved through adding ultra-thin white topping on asphalt road, which is a Portland concrete overlay on existing asphalt to increase asphalt albedo to 42%. Materials that are made from albedo in buildings' facades and roofs were also increased to 50% and interlock to 40%.

8.7 Scenarios' Results

8.7.1 Assessing Climate Change

This section presents the thematic maps extracted from ENVI-met simulation for the following five factors, dry bulb temperature (DBT), wind speed (WS), relative humidity (RH), mean radiant temperature (MRT) and predicted mean vote (PMV), to assess climate change impact in urban areas in 2080 scenario (Fig. 8.6) and compare it to that of 2016. The maps are extracted at three different hours of the day (08:00, 13:00 and 17:00). The different climatic parameters extracted from the

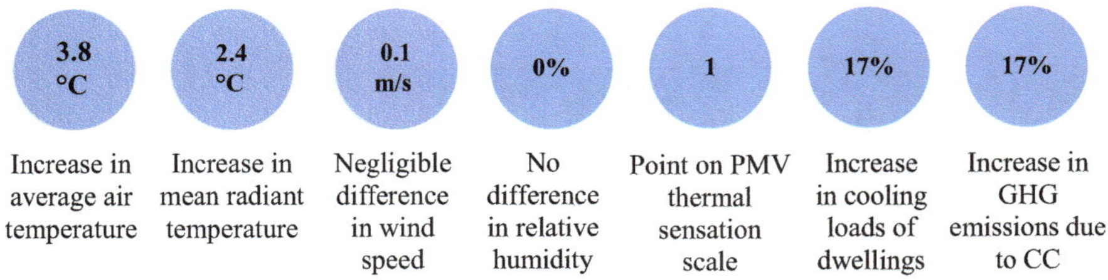

Fig. 8.6 The housing project 'Asmarat' by 2080 overview

thematic maps (Figs. 8.7, 8.8, 8.9, 8.10, and 8.11) were compared to determine climate change impacts.

The simulation study indicates that there is a significant increase in air temperature by 3.86 °C due to the climate change and average temperature increases from 33.80 °C to 37.66 °C. Relative humidity is at the same value in both models of the Asmarat As-built 2016 and 2080. As shown in Fig. 8.14, there is slight decrease in average wind speed of Asmarat from 1.19 m/s (As-built 2016) to 1.11 m/s (As-built 2080) (7%). The average mean radiant temperature (MRT) increases from 66.51 in the Asmarat As-built 2016 model to 68.95 in that of 2080. The PMV thermal comfort indices indicate that all simulated models are outside the thermal comfort zone, which have PMV scale ranging between −0.5 and +0.5. The resulting PMV index (Fig. 8.16) illustrates that climate change increases the intensity of warm thermal discomfort throughout the day. The PMV value of the Asmarat As-built 2016 model has increased from 5.42 to 6.40, which is similar to the value of the 2080 model. A summary of the climate change impacts in Asmarat, Egypt, by 2080 is listed in Fig. 8.6.

Box 8.1 Climate Change

Future anthropogenic climate change is likely to increase the average air temperature by 3.81 °C by 2080 to be followed by an increase in mean radiant temperature by approximately 2.44 °C.

Box 8.2 Adaptation Scenario Importance

Adaptation through buildings' retrofitting and cities' upgraded strategies plays a vital role in coping with global warming through enhancing outdoor thermal comfort. Hence, it's recommended to incorporate the four retrofitting proposal.

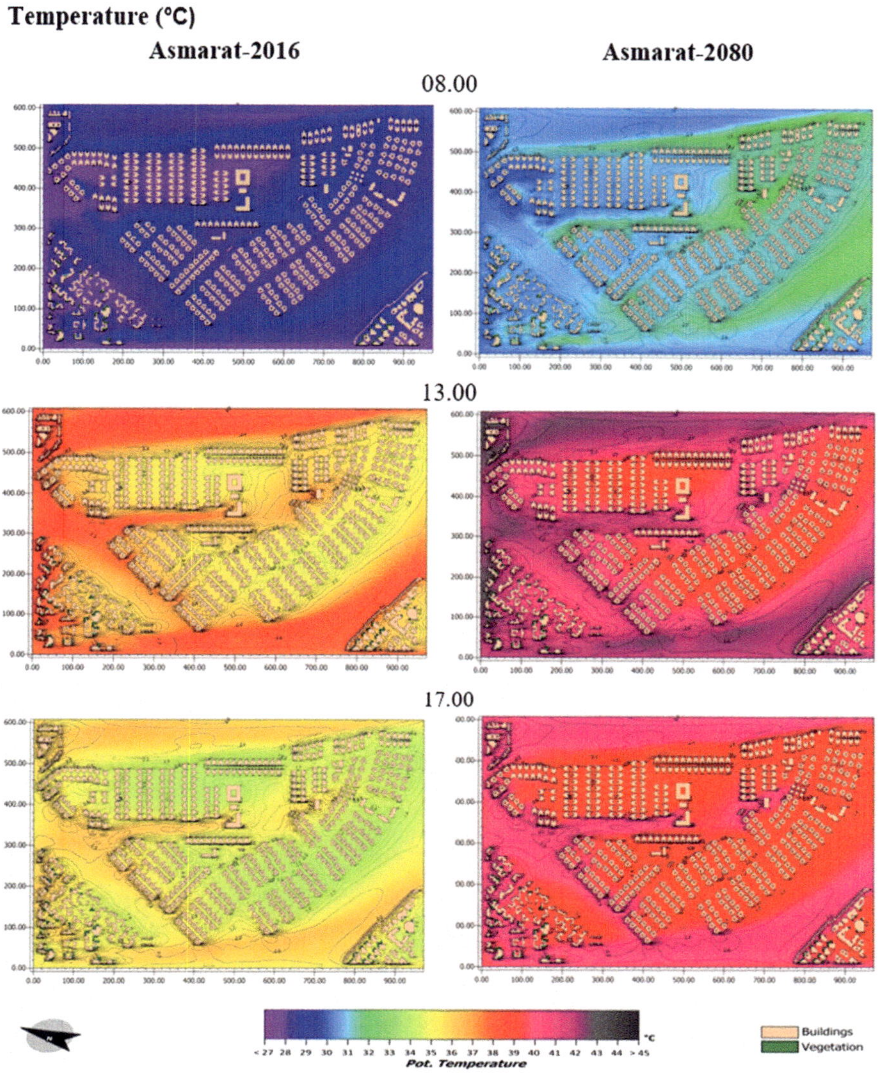

Fig. 8.7 Temperature maps of Asmarat 2016 and 2080 models

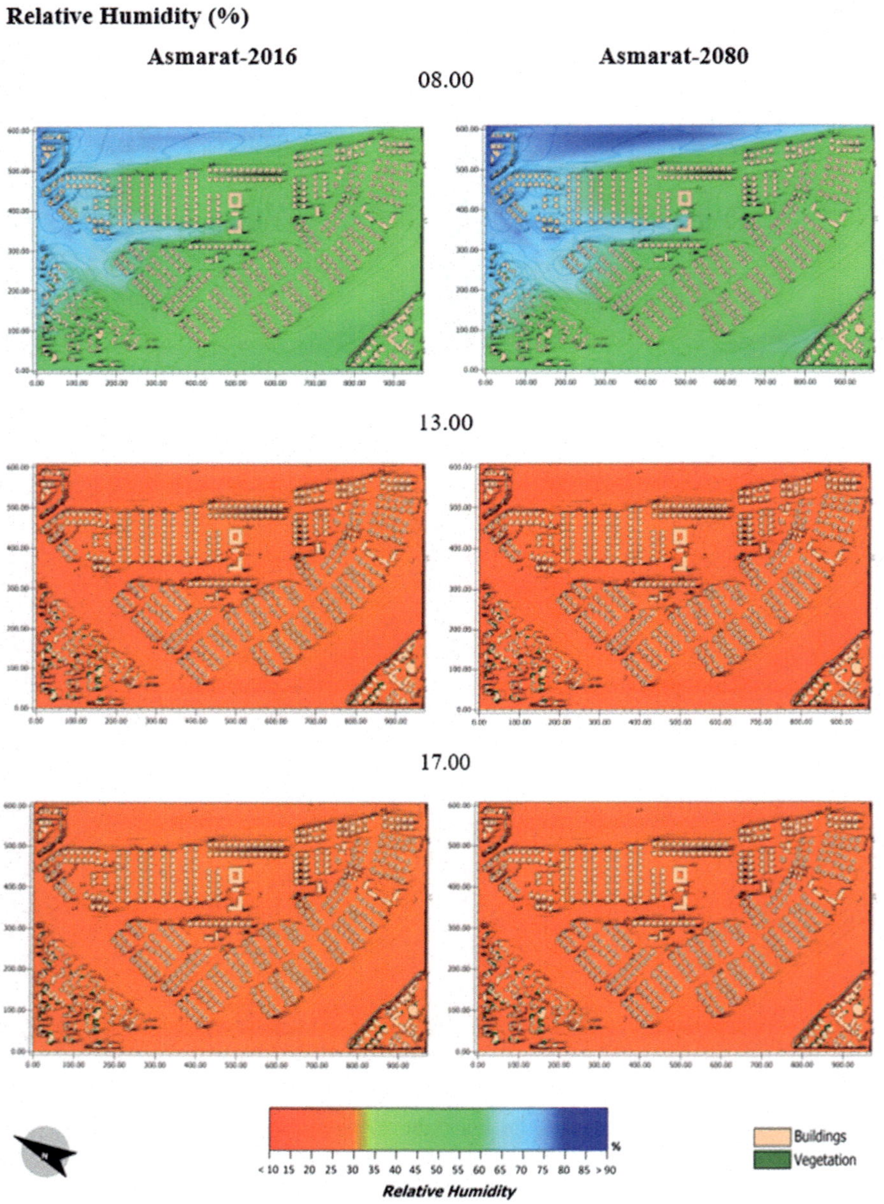

Fig. 8.8 Relative humidity maps of Asmarat 2016 and 2080 models

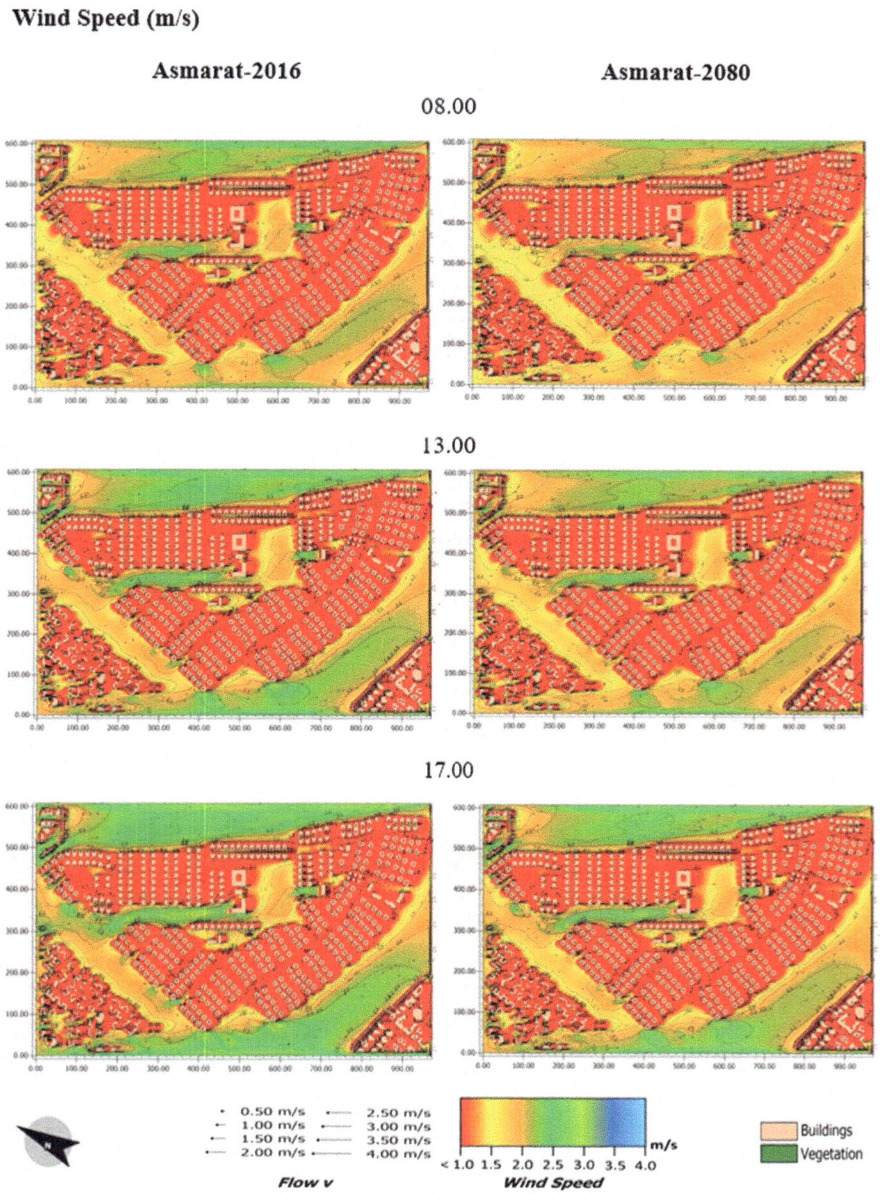

Fig. 8.9 Wind speed maps of Asmarat 2016 and 2080 models

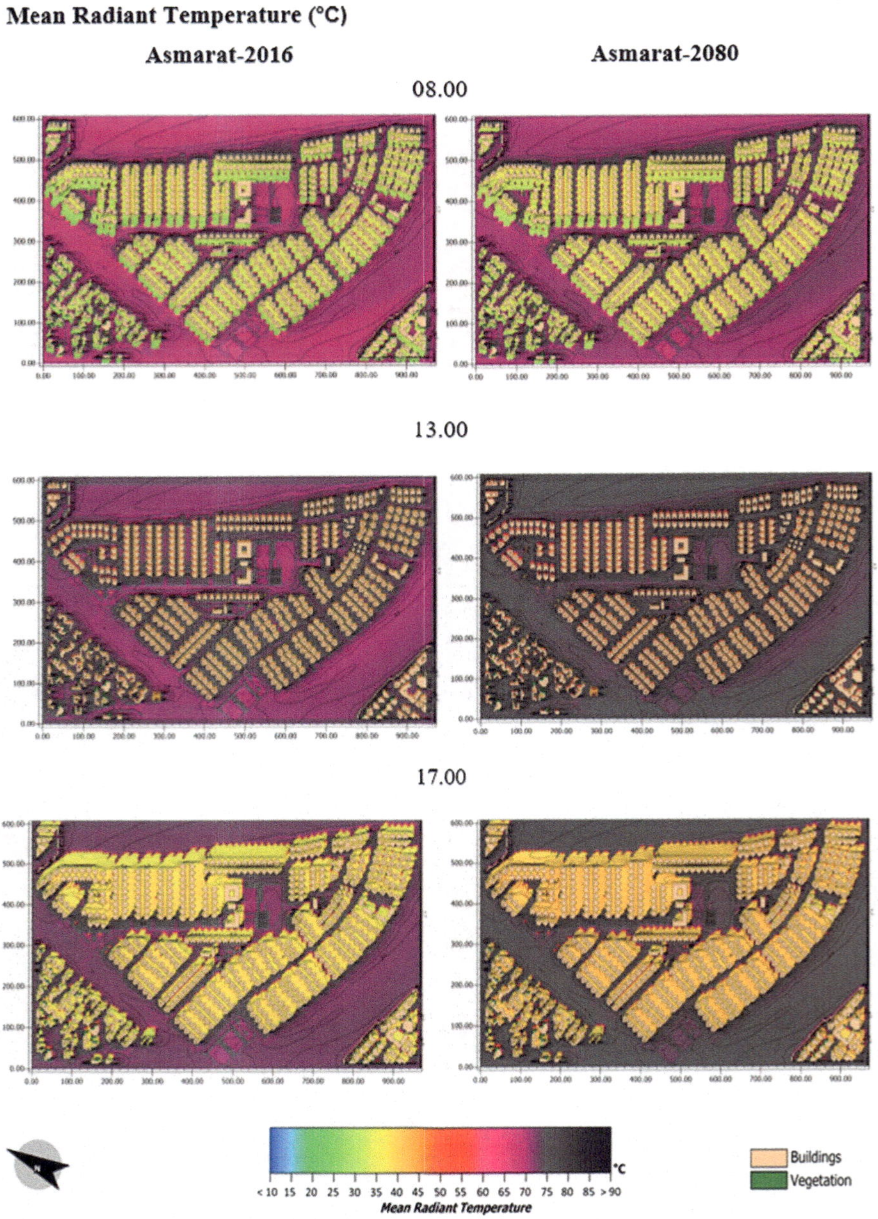

Fig. 8.10 MRT maps of Asmarat 2016 and 2080 models

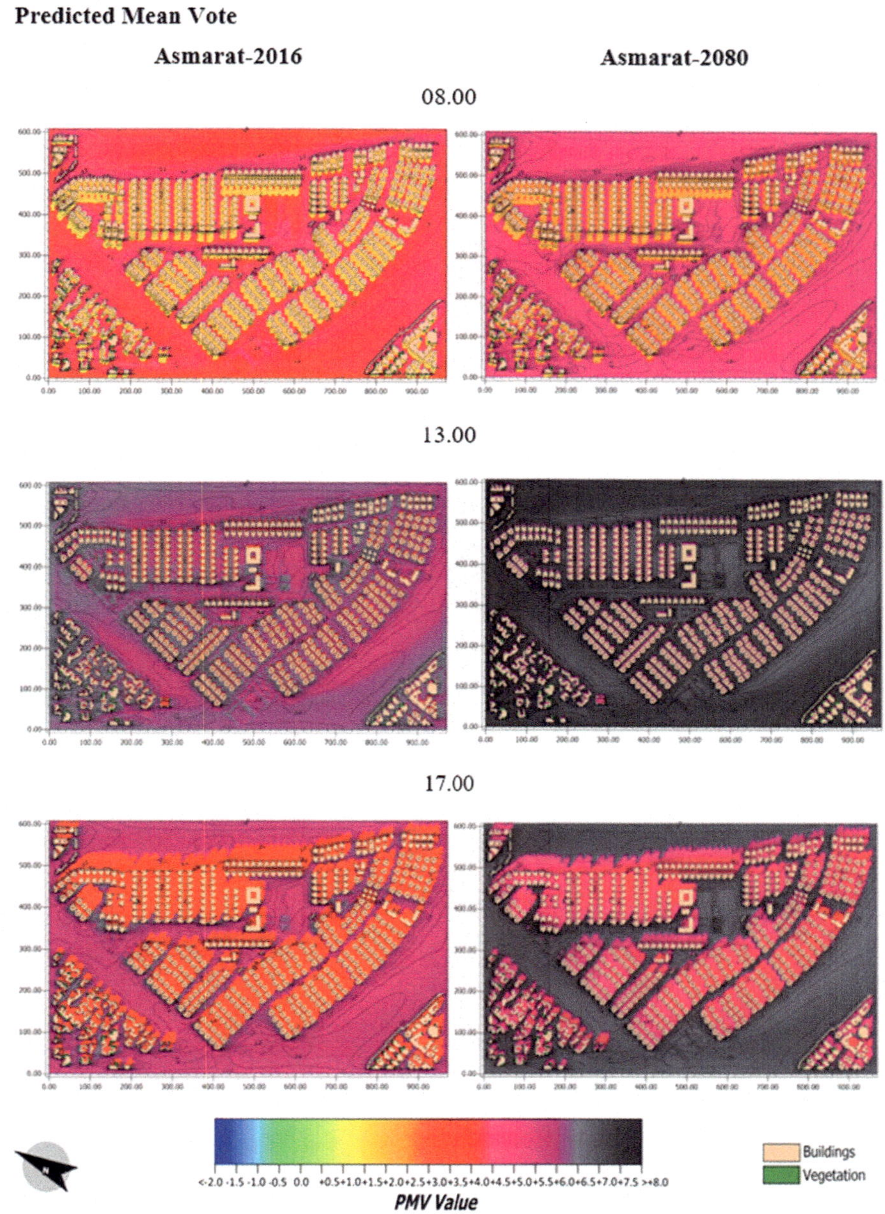

Fig. 8.11 PMV maps of Asmarat 2016 and 2080 models

8.7.2 *Testing Passive Cooling Design Configurations for Adaptation*

The following tables (Tables 8.3, 8.4, 8.5, 8.6, and 8.7) show the extracted weather data matrices' output for the climatic parameters (air temperature, relative humidity, wind speed, mean radiant temperature – MRT, and predicted mean

Table 8.3 Air temperature data output

Time	As-built 2016	As-built 2080	Envelop	Waterfeatures	Elevation	Vegetation	MTLalbedo	Orientation
8	28.43	31.17	31.06	31.15	30.33	30.76	31.02	30.92
9	30.31	33.85	33.75	33.81	33.50	33.58	33.61	33.92
10	31.55	35.11	35.02	35.06	34.85	34.72	34.83	35.14
11	32.99	36.67	36.57	36.61	36.35	36.19	36.39	36.81
12	34.76	38.64	38.55	38.58	38.28	38.05	38.35	38.75
13	36.07	40.12	40.04	40.06	39.83	39.53	39.84	40.08
14	36.59	40.79	40.72	40.73	40.56	40.21	40.54	40.65
15	36.46	40.75	40.68	40.70	40.53	40.22	40.55	40.64
16	35.90	40.23	40.18	40.18	40.00	39.71	40.09	40.12
17	34.92	39.31	39.27	39.26	39.10	38.82	39.23	39.22
AVG	33.80	37.66	37.58	37.61	37.33	37.18	37.45	37.63

Table 8.4 Relative humidity data output

Time	As-built 2016	As-built 2080	Envelop	Waterfeatures	Elevation	Vegetation	MTLalbedo	Orientation
8	55	57	57	57	66	58	57	54
9	44	45	45	45	49	47	45	45
10	39	39	39	39	40	40	39	38
11	35	34	34	34	36	35	35	34
12	31	30	30	30	31	31	30	30
13	28	27	27	27	28	28	27	27
14	27	26	26	26	26	27	26	26
15	27	26	26	26	26	26	26	26
16	27	26	26	26	27	27	26	26
17	29	27	27	27	28	28	27	27
AVG	34%	34%	34%	34%	36%	35%	34%	33%

Table 8.5 Wind speed data output

Time	As-built 2016	As-built 2080	Envelop	Waterfeatures	Elevation	Vegetation	MTLalbedo	Orientation
8	1.12	1.07	1.07	1.07	1.17	0.96	1.07	1.05
9	1.11	1.04	1.04	1.05	1.15	0.93	1.04	1.03
10	1.14	1.06	1.06	1.06	1.14	0.91	1.05	1.03
11	1.16	1.09	1.08	1.09	1.16	0.97	1.08	1.05
12	1.18	1.11	1.10	1.11	1.16	0.99	1.10	1.06
13	1.20	1.12	1.12	1.12	1.17	1.01	1.12	1.08
14	1.22	1.13	1.13	1.13	1.18	1.02	1.13	1.09
15	1.23	1.14	1.14	1.14	1.19	1.03	1.14	1.10
16	1.25	1.16	1.15	1.16	1.20	1.04	1.15	1.12
17	1.26	1.17	1.17	1.17	1.21	1.05	1.16	1.13
AVG	1.19	1.11	1.11	1.11	1.17	0.99	1.10	1.07

Table 8.6 Mean radiant temperature data output

Time	As-built 2016	As-built 2080	Envelop	Waterfeatures	Elevation	Vegetation	MTLalbedo	Orientation
8	56.45	57.79	57.75	57.77	55.31	57.14	63.32	56.70
9	66.84	68.43	68.41	68.42	66.57	65.58	75.16	64.04
10	67.85	69.87	69.84	69.85	69.78	65.49	77.06	65.14
11	68.17	70.58	70.55	70.55	70.44	67.54	78.45	72.37
12	67.64	70.39	70.36	70.37	70.39	67.59	78.55	70.97
13	73.49	76.23	76.20	76.20	76.34	73.07	83.91	75.41
14	74.27	77.04	77.02	77.02	76.73	74.10	83.58	79.05
15	72.99	75.76	75.73	75.74	75.32	73.21	81.61	79.15
16	62.97	65.93	65.92	65.92	65.72	64.15	70.42	72.78
17	54.44	57.45	57.43	57.42	55.47	54.71	61.03	60.83
AVG	66.51	68.95	68.92	68.93	68.21	66.26	75.31	69.65

Table 8.7 Predicted mean vote data output

Time	As-built 2016	As-built 2080	Envelop	Waterfeatures	Elevation	Vegetation	MTLalbedo	Orientation
8	3.71	4.39	4.37	4.39	4.05	4.32	4.82	4.24
9	4.87	5.74	5.72	5.73	5.50	5.51	6.32	5.38
10	5.15	6.05	6.03	6.04	5.97	5.66	6.67	5.65
11	5.41	6.36	6.34	6.34	6.28	6.05	7.05	6.56
12	5.67	6.67	6.66	6.66	6.60	6.36	7.37	6.76
13	6.40	7.45	7.44	7.44	7.40	7.09	8.14	7.38
14	6.56	7.64	7.63	7.63	7.55	7.30	8.22	7.82
15	6.41	7.51	7.50	7.50	7.42	7.21	8.03	7.82
16	5.44	6.54	6.53	6.53	6.48	6.31	6.90	7.12
17	4.59	5.67	5.66	5.66	5.49	5.38	5.95	5.93
AVG	5.42	6.40	6.39	6.39	6.27	6.12	6.95	6.47

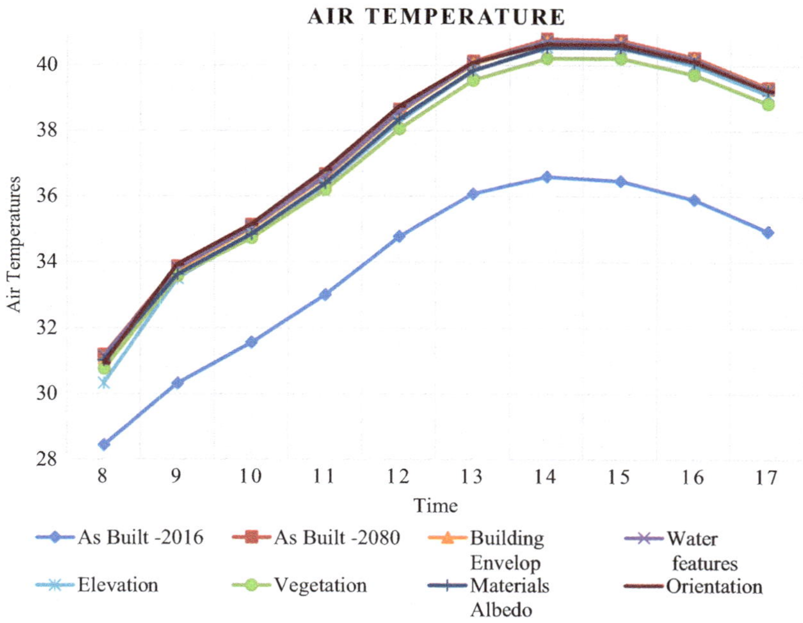

Fig. 8.12 Air temperature data output

vote – PMV) applied for the as built 2016 and the as built model scenario 2080. Also Figs. 8.12, 8.13, 8.14, 8.15, and 8.16 present the line charts for the studied climatic parameters that compare the differences between the alternative models, which encompass six configurations: buildings' envelop, water feature, buildings' elevation, vegetation, high albedo and orientation against the as-built 2016 and the as-built model scenario 2080.

The different climatic parameters (numbered from 'a' till 'd') that were extracted from the thematic maps and graphs were compared to determine proposal performance in terms of the environmental factors that directly affect human thermal comfort (referred to as 'e'). The results are as follows:

(a) *Air Temperatures*

The simulation study shows that there is a significant increase in air temperature by 3.9 °C due to the climate change; average temperature increases from 33.8 °C to 37.7 °C. In tested retrofitting proposals, the differences between air temperatures in the four proposals, buildings' envelop, water features, material albedo and orientation in addition to As-built 2080 case, weren't significant, approximately averaging 37.6 °C, whereas the vegetation proposal has the lowest air temperature (average of 37.2 °C) followed by buildings' elevation proposal (average of 37.3 °C) during that time (Fig. 8.12).

(b) *Relative Humidity*

Relative humidity is at the same value in both models of Asmarat As-built 2016 and 2080. The buildings' elevation proposal has the highest relative humidity ratio

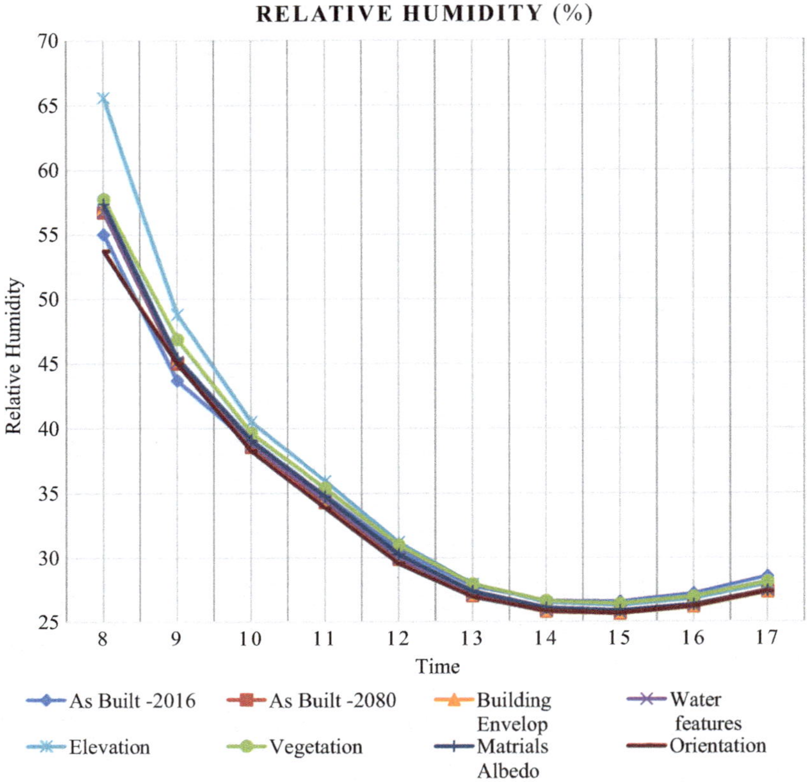

Fig. 8.13 Relative humidity data output

of 36% followed by that of vegetation proposal (35%), while the proposals of buildings' envelop, water features and material albedo in addition to As-built 2080 case have a ratio of 34% as that of Asmarat 2016, while orientation proposal relative humidity ratio is 33% (Fig. 8.13).

(c) *Wind Speed*

As shown in Fig. 8.14, there is slight decrease in average wind speed of Asmarat from 1.19 m/s (As-built 2016) to 1.11 m/s (As-built 2080) (7%). Results indicate that buildings' elevation proposal has the highest wind speed average of 1.17 m/s, and the other proposals such as buildings' envelop, water features and material albedo approximately equal to that of the As-built 2080 are recorded with an average of 1.11 m/s, followed by the orientation proposal (1.1 m/s). Vegetation proposal has the lowest wind speed of 1.0 m/s.

(d) *Mean Radiant Temperature*

The average mean radiant temperature (MRT) increases from 66.51 in the Asmarat As-built 2016 model to 68.95 in that of 2080. The vegetation proposal has

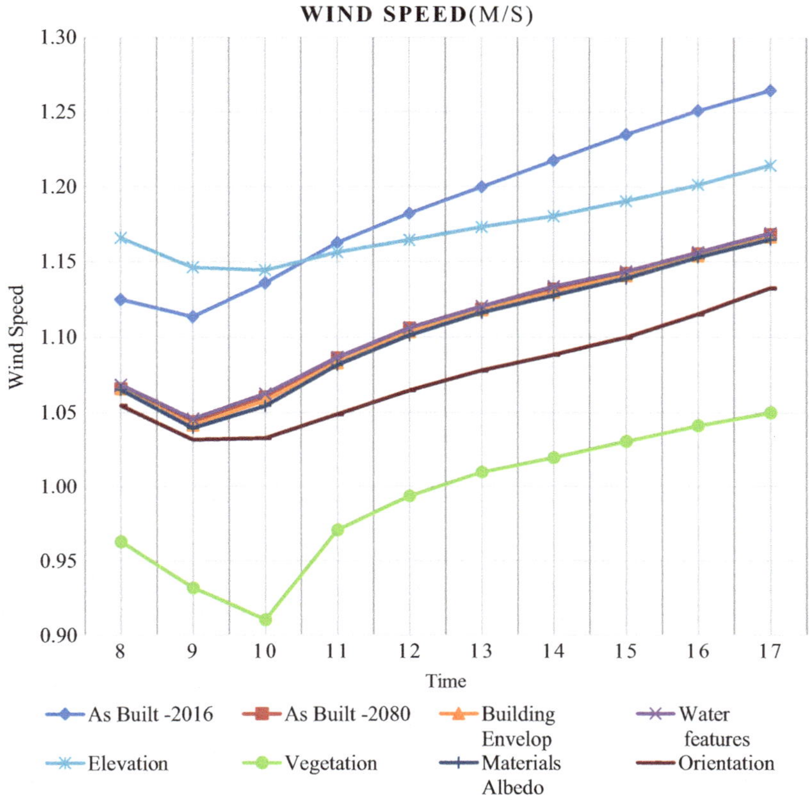

Fig. 8.14 Wind speed data output

the lowest MRT (66 °C) followed by that of elevation proposal (68 °C), while the rest of the proposals are approximately at the same MRT (69 °C) but the material albedo proposal has the highest MRT (75 °C) as presented in Fig. 8.15.

(e) *Predicted Mean Vote*

The PMV thermal comfort indices indicate that all simulated models are outside the thermal comfort zone, which have PMV scale ranging between −0.5 and +0.5. The resulting PMV index (Fig. 8.16) shows that climate change increases the intensity of warm thermal discomfort throughout the day. The PMV value of the Asmarat As-built 2016 model has increased from 5.42 to 6.40, which is similar to the value of the 2080 model. In terms of passive cooling models' proposals, the vegetation proposal has the lowest PMV value (6.12) followed by elevation proposal (6.27), while PMV of buildings' envelop and water feature proposal is 6.39 approximately, which is the same of As-built 2080. Orientation proposal is slightly higher than As-built 2080, but material albedo proposal has the highest PMV value (6.95).

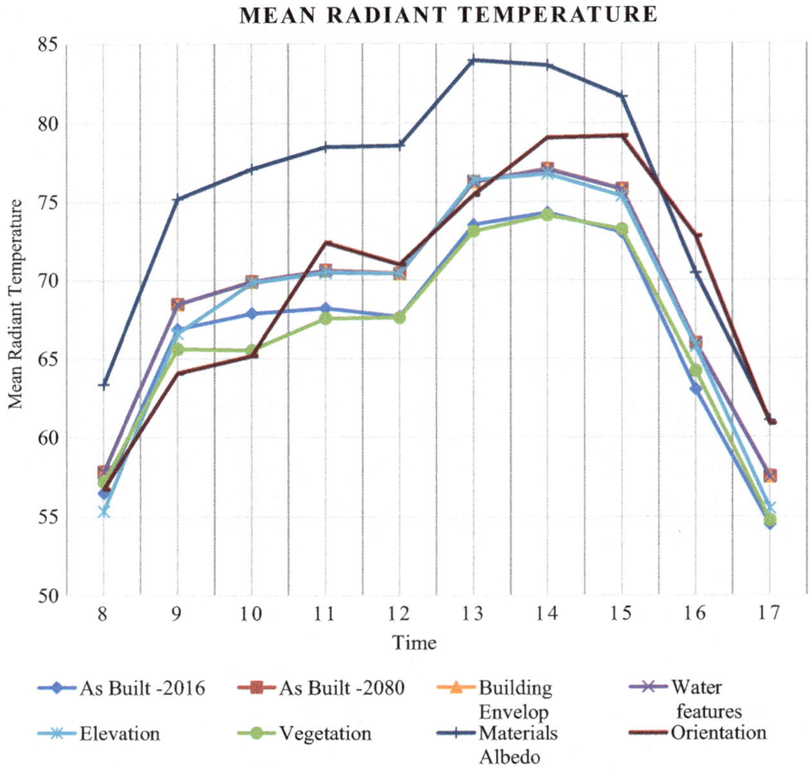

Fig. 8.15 Mean radiant temperature data output

8.7.2.1 Cooling Load

Using design builder software and coupling simulation, conclusions regarding the building cooling can be drawn (Fig. 8.17). A significant increase in cooling loads of the Asmarat project was found from 82 kW to 97 kW in the As-built 2016 and the As-built 2080 respectively. The buildings' envelop proposal has the lowest cooling load, which is 70 kW. Also, the cooling loads of the material albedo proposal can be considered low, followed by that of the orientation, vegetation, and water features, which has the lowest cooling load values, even lower than that of the Asmarat (As-built 2080). Finally, the buildings' elevation proposal has the highest cooling load, which is 108 kW. The increase in the cooling load of the buildings' elevation proposal is considered significant compared to the other configurations.

8.7.2.2 GHG Footprint

There is a significant increase in greenhouse gas (GHG) emissions by 0.01 metric tons in 2080 (Fig. 8.18). The buildings' envelop proposal has the lowest GHG footprint with 0.05 metric tons. The models of material albedo proposal followed by

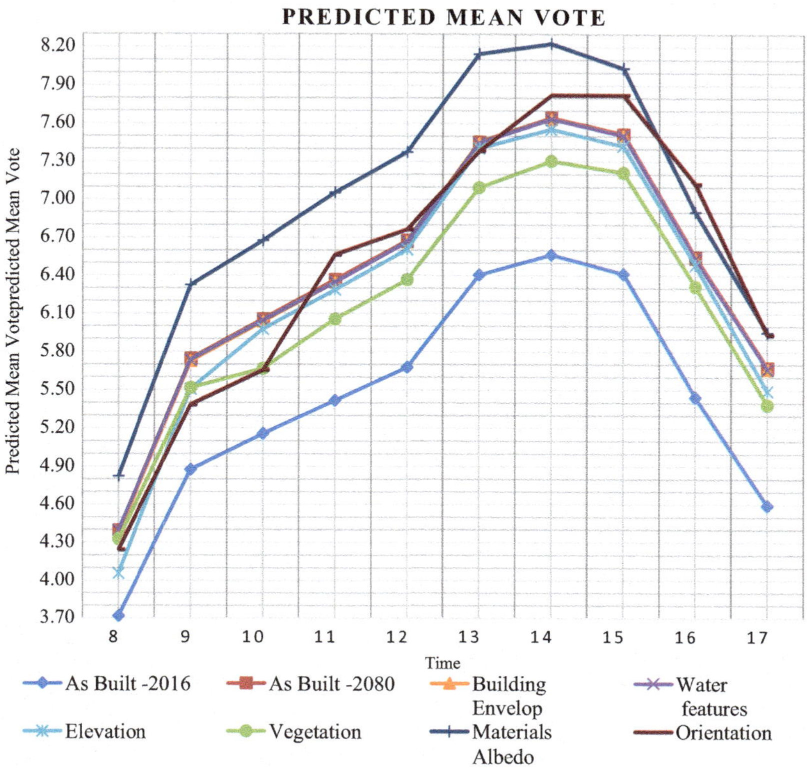

Fig. 8.16 Predicted mean vote data output

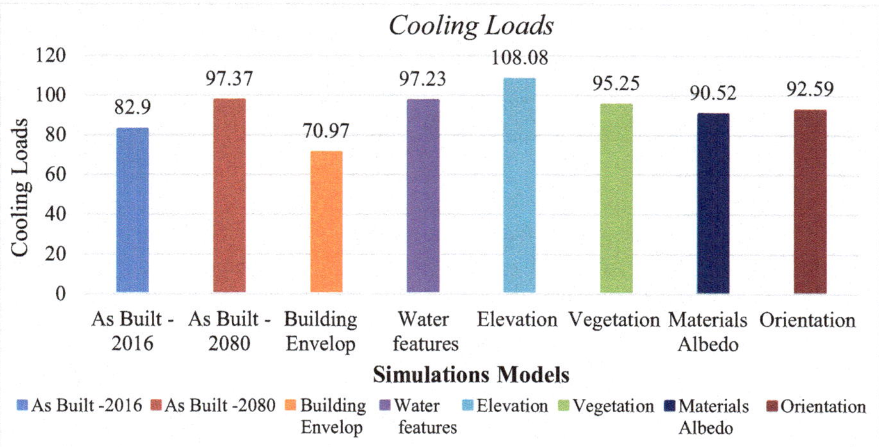

Fig. 8.17 Cooling loads of simulation models

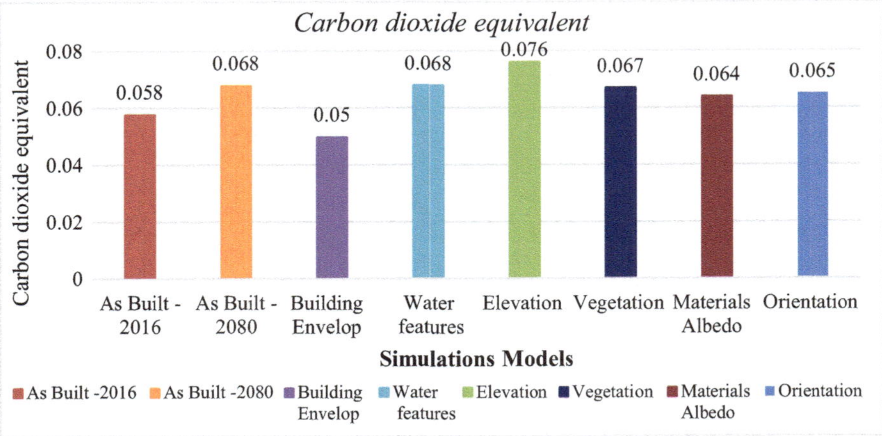

Fig. 8.18 Carbon dioxide (CO_2) equivalent of simulation models

Table 8.8 Coupling Results

Models	Compared to	PMV	Delta (PMV)	Ratio of delta %	Cooling loads	Delta (cooling loads)	Ratio of delta %	GHG emission	Delta (GHG)	Ratio of delta %
As-built 2016	–	5.42	–	–	82.9	–	–	0.058	–	–
As-built 2080	As-built 2016	6.4	0.98	18	97.37	14.47	17	0.068	0.01	17
Buildings elevation 2080	As-built 2080	6.27	−0.13	−2	108.08	10.71	11	0.076	0.008	12
Buildings envelop 2080	As-built 2080	6.39	−0.01	−0.2	70.97	−26.4	−27	0.05	−0.018	−26
Vegetation 2080	As-built 2080	6.12	−0.28	−4	95.25	−2.12	−2	0.067	−0.001	−1
Water features 2080	As-built 2080	6.39	−0.01	−0.2	97.23	−0.14	−0.1	0.068	0	0
Orientation 2080	As-built 2080	6.47	0.07	1	92.59	−4.78	−5	0.065	−0.003	−4
High albedo 2080	As-built 2080	6.95	0.55	9	90.52	−6.85	−7	0.064	−0.004	−6

Negative impact | Positive impact

orientation, then vegetation and water features have a lower GHG than that of the Asmarat As-built 2080 model. The buildings' elevation proposal has the highest GHG footprint of 0.076 metric tons.

8.7.2.3 Coupling Results

Table 8.8 presents coupling results from ENVI-met and DesignBuilder software for the base case in 2016 and 2018 along with the climate change adaptation scenarios.

Box 8.3 Simulation Scenarios

- *Vegetation* has the best performance on PMV due to cooling effects through evaporation and shading. These are the main processes through vegetation which affects climate; it is ranked the fourth in indoor comfort.
- *Raised buildings' elevation* is the second in performance on PMV scale due to the continuity of air movement that improves the cooling sensation and reducing MRT; it has the worst performance in indoor due to increase in heat gain of ground floor.
- *Buildings' envelopes* impact on outdoor is negligible because it acts as a thermal barrier regulating interior temperatures. It may have a slightly positive effect on outdoor thermal comfort due to decreased storage of sensible heat in the construction materials; it is the best proposal in indoor performance.
- *Water feature* impact on outdoor is negligible due to water bonds that have little effect on temperature due to thermal stratification phenomenon that refers to high temperature of pond top than bottom where the sun warms the surface water unlike the cooling effect of fountains; it is the fifth for indoor comfort.
- *Orientation* is slightly negative, despite the reduction in air temperature. It has a significant increase in MRT due to the decrease in buildings' shading; it is the third in indoor performance due to reduction in heat gain.
- *High material albedo* is the worst proposal in outdoor despite its air temperature reduction. However, it encountered with a significant increase in MRT due to reflected short wavelength radiation by high albedo. It is the second in indoor as the reflected radiation reduces heat gain.

8.8 Conclusions

The assessment presents the outcomes of the climate change adaptation scenarios 2016 and 2080, which is based on the analytical models and simulations.

It is clear that the future anthropogenic climate change is likely to increase the average air temperature by 3.81 °C by 2080. This will be followed by an increase in the mean radiant temperature (MRT) approximately at 2.44 °C with a slight difference – almost negligible – in the average wind speed (approximately equals 0.1 m/s), whereas the simulations of relative humidity indicate that there is no difference in this parameter. Changing of climatic parameters will result in increased magnitude of heat stress and thermal discomfort throughout the set day, approximately 1 point on PMV thermal sensation scale resulting in 17% increase in cooling loads and GHG emissions of Asmarat. In terms of the six passive cooling models' proposals, results indicate that the vegetation has the most positive impact on thermal comfort, followed by raised buildings' elevation. However, the buildings' envelop and water

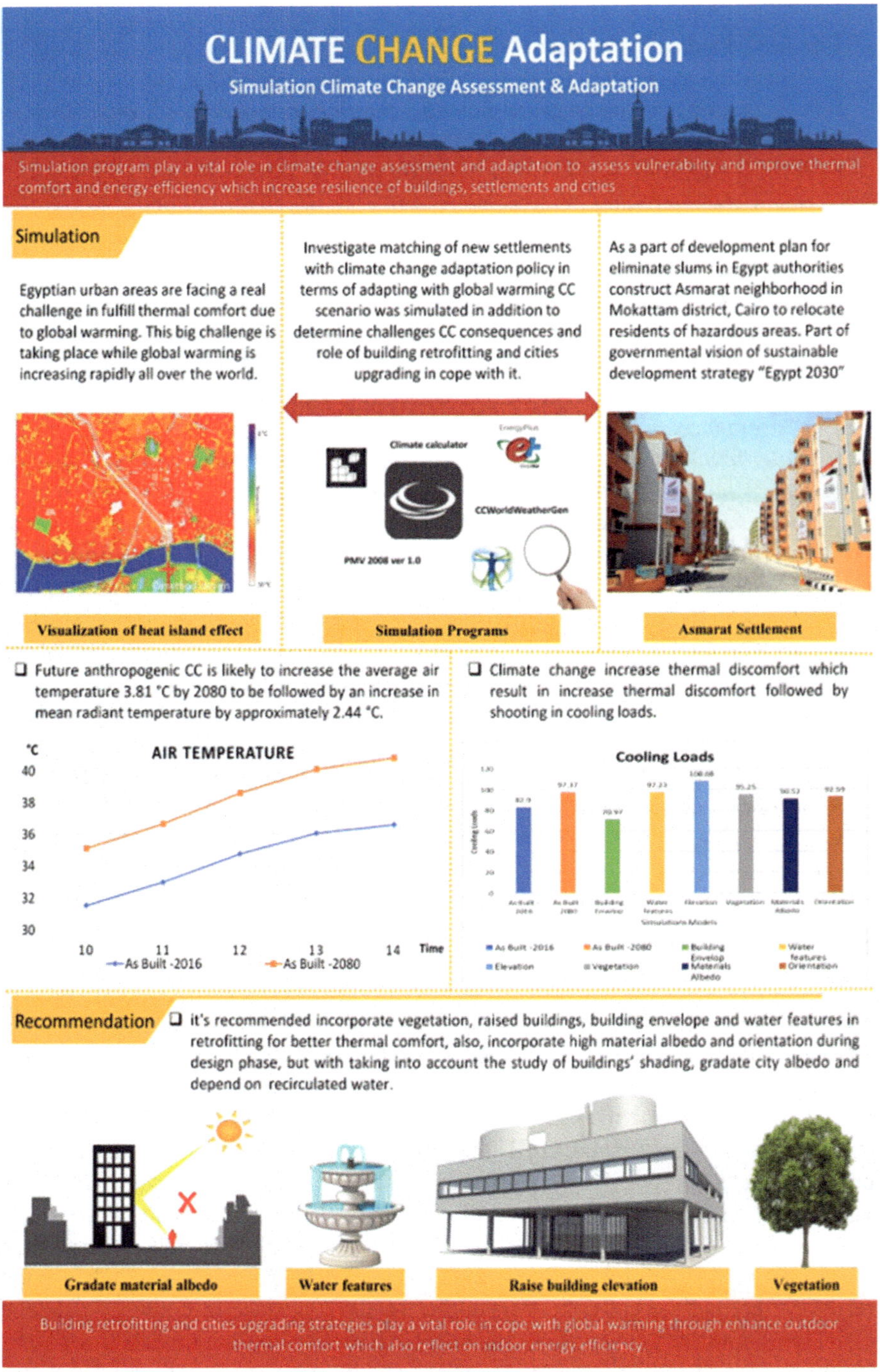

Fig. 8.19 Infographic depicting the outlines of this chapter. (Source: Developed by authors)

features have very little impact, whereas the orientation proposal has slightly negative impact followed by a large margin in the high material albedo proposal that has the worst performance impact on the PMV scale. Vegetation has the best performance on the PMV scale due to cooling effect through evaporation and shading. These are the main processes through which vegetation affects climate; hence it is ranked the fourth in indoor thermal comfort performance due to the reduction in temperature and heat gain accordingly. Raised buildings' elevation model is ranked second in terms of performance on PMV scale due to the continuity of air movement that improves cooling sensation and reduces MRT. It has the worst performance in indoor due to increase in heat gain of ground floors. In addition, buildings' envelop impact on outdoor is negligible because the main role of buildings' envelop is to act as a thermal barrier regulating interior temperatures. It may have a slightly positive effect on the outdoor thermal comfort due to decreased storage of sensible heat in the construction materials which is one of urban heat island reasons. Nevertheless, it is the best proposal in terms of indoor performance due to insulations that reduce heat gain and cooling loads. In terms of water features, the impact on outdoor is negligible because the water bonds have little effect on temperature due to thermal stratification phenomenon that refers to high temperature of pond top than bottom. The water feature proposal is ranked the fifth for indoor thermal comfort performance. Orientation is slightly negative despite proposals to reduce air temperature, but the reduction has a significant increase in MRT due to the decrease in buildings' shading. It is ranked the third in indoor performance. High material albedo (HMA) is the worst proposal on outdoor thermal comfort despite its reduced air temperature. However, the reduction has a significant increase in MRT due to reflected short wavelength radiation by HMA. Thus the thermal comfort index goes warmer. It is ranked the second in indoor performance because of the reflected radiation which reduces heat gain.

Therefore, the climate change scenario 2080 results suggest incorporating vegetation, raised buildings, buildings' envelop and water features to enhance urban areas for better thermal comfort. Also, it recommends incorporating HMA and orientation during design phase and taking into account the impact of buildings' shading when defining city orientation to maximize shading exclusion using high albedo materials on pavement to prevent the negative impact of reflected short wavelength radiation on pedestrians. In addition, low albedo is to be applied at the bottom part whereas high albedo at the top of the building walls and roofs for lower impact on pedestrians' thermal comfort. It is also recommended to insulate ground floor when using raised buildings' elevation strategy to reduce heat gain of ground floor for decreasing indoor cooling loads. Finally, climate change simulation models have proven to be effective tools in predicating climate change adaptation scenarios in developing countries. A summary of climate change adaptation using simulation and the results of the chapter (info-graphic chart) is shown in Fig. 8.19.

References

1. Holli Riebeek, "Global Warming", NASA Earth Observatory's, 2007.
2. Fathy El ashmawy and Ahmad abdelaal, "Characteristics of heat waves over Egypt and east Mediterranean", in WMO technical conference on climate services building on clips legacy, Heidelberg, Germany, 30 June–2 July 2014.
3. "Egypt Third National Communication Under the United Nations Framework Convention on Climate Change", Egyptian Environmental Affairs Agency (EEAA), March 2016.

Index

M. M. Aboulnaga et al., *Urban Climate Change Adaptation in Developing Countries*, https://doi.org/10.1007/978-3-030-05405-2

Printed by Printforce, the Netherlands